SCHELLING'S PRACTICE OF THE WILD

SUNY series in Contemporary Continental Philosophy
Dennis J. Schmidt, editor

Schelling's Practice of the Wild

Time, Art, Imagination

JASON M. WIRTH

Published by State University of New York Press, Albany

Printed in the United States of America

For information, contact State University of New York Press, Albany, NY
www.sunypress.edu

Production, Diane Ganeles
Marketing, Anne M. Valentine

Library of Congress Cataloging-in-Publication Data

Wirth, Jason M., 1963-
Schelling's practice of the wild : time, art, imagination / Jason M. Wirth.
pages cm. — (SUNY series in contemporary continental philosophy)
Includes bibliographical references and index.
ISBN 978-1-4384-5679-9 (hardcover : alk. paper)
ISBN 978-1-4384-5678-2 (paperback : alk. paper) — ISBN 978-1-4384-5680-5 (e-book) 1. Schelling, Friedrich Wilhelm Joseph von, 1775–1854. 2. Philosophy of nature. 3. Thought and thinking. 4. Aesthetics. I. Title.
B2899.N3W57 2015
113—dc23

2014027723

10 9 8 7 6 5 4 3 2 1

To Elizabeth Myoen Sikes
in gratitude for sharing a practice of the wild

None of our spiritual thoughts transcends the earth.

—Schelling to Eschenmayer (1812) (I/8, 169)

Dao, the way of Great Nature: eluding analysis, beyond categories, self-organizing, self-informing, playful, surprising, impermanent, insubstantial, independent, complete, orderly, unmediated . . .

—Gary Snyder, *The Practice of the Wild* (10)

CONTENTS

PREFACE

The last two decades have enjoyed a considerable renaissance and reappraisal of Schelling's remarkable body of philosophical work. In addition to the laudable feat of bringing the long neglected works of Schelling to our philosophical attention, this shift in the philosophical terrain allows for new kinds of work to appear. Now that the basic case for both Schelling's intrinsic interest and contemporary relevance has been made, there is an opportunity to move beyond issuing reports on Schelling and merely explicating these admittedly difficult texts. We can now think with and through Schelling, accompanying him as an opportunity to explore and develop the fundamental issues at stake in his thought. One need not restrict oneself to reportage—even of the most hermeneutically savvy kind. One can also appreciate the questions that Schelling often developed in such startling and original ways as philosophical problems that are worthy in their own right. Moreover, one can insist that these are commanding issues, problems that speak to some of the great matters of human living and dying.

Following a range of remarkable texts to have appeared in Germany and France and elsewhere since the middle of the twentieth century, some bold and original new studies of Schelling have appeared in English, including the works of Iain Hamilton Grant, Bruce Matthews, Sean McGrath, Tilottoma Rajan, Markus Gabriel, Bernard Freydberg, and many others.[1] It is to this burgeoning tradition that I aspire with this book.

Continental philosophy has to some extent earned the bad rap of becoming a kind of philosophical ventriloquism, obsequiously satisfied with becoming a puppet for the voice about which one writes. There are increasingly many inspiring counterexamples to this practice, and it is this latter kind of company that this book desires to keep. This book is not a report about a philosopher. It seeks to engage Schelling the philosopher *philosophically* and its line of inquiry is the problem of the *relationship of time and the imagination*, both in Schelling's remarkable thought and as a worthy problem in its own right. I call this relationship, both in terms of its concepts and the modes of human living that it endears, *Schelling's practice of the wild*. It is a problem that bears on the questions of nature, art, philosophical religion (mythology and revelation), and history.

Wild Nature and Its Practice

Schelling argued, in as many different ways as he could throughout his long philosophical career, that nature is alive. Famously in the *Freedom* essay, Schelling charged that the prevailing modern view of natural science with its positivistic representation of nature, or more precisely, its view of nature as representable, is nature-cide, the fatal flaw that epitomizes thinking in the wake of the Enlightenment: "nature is not present to it" for modernity "lacks a living ground [*die Natur für sich nicht vorhanden ist, und daß es ihr am lebendigen Grunde fehlt*]" (I/7, 361). Nature therefore becomes a vague abstraction and its forces become mere repetitions of the same (so-called natural laws). Elsewhere Schelling speaks of the "the true annihilation [*Vernichtung*] of nature" (I/5, 275) as it becomes a dead object under the control and interests of the human subject. In his early *Naturphilosophie*, Schelling creatively retrieved the ancient doctrine of the *anima mundi*, living nature, animated by its ψυχή or *anima*, what Plato in the *Timaeus* dubbed the ψυχή κόσμου, the soul of the cosmos, or what Schelling called *die Weltseele*, the soul of the world. There is good reason for the natural sciences to be suspicious of such descriptions. Life is reserved for the things that are alive, not for rocks and houses. Moreover, the analogy between a living being and a vital universe is quite loose and doubtless falls short in many respects.

This, however, is already Schelling's point. To say that the universe is alive is not to say that it is a living thing, the superorganism that contains all other organisms.[2] The life of nature holds together organically the living and dying of living things and the coming and going of nonliving things. This should not be confused with some *élan vital* or vitalism or life force (*Lebenskraft*). Life is not external to the living, coming into the living from afar. As we see in the remarkable essay *Über das Verhältniß des Realen und Idealen in der Natur, oder Entwickelung der ersten Grundsätze der Naturphilosophie an den Principien der Schwere und des Lichts*, written in 1806 to complete *Von der Weltseele*:

> It is undeniable that apart from the external life of things, an inner life is manifest, whereby they are capable of sympathy and antipathy and of the perception of other, not even immediately present things; it is also undeniable that the universal life of things is at the same time the particular life of the individual.
>
> Since this is the principle by which the infinity of things is posited universally as eternity and present, so it is equally that which forms the permanent in time, individual circuits, as it were, in the all-enclosing circle of eternity, that adorn the years, months and days; and must we not agree with Plato in naming this all-organizing and improving principle of global and universal wisdom and the majestic soul of the all? (I/2, 371)[3]

Life is the sovereign presentation of nature, its free coming into form and necessity. Schelling's turn here to the *Timaeus* and the χώρα at the heart of nature marks a different conception of life, a way of life that I am marking by the word *wild* that I am borrowing from the American poet Gary Snyder. *Nature is through and through alive because it is wild.*

Although I discuss this in more detail in my opening chapter, we can say that the wild is not the mere opposite of the tame and the ordered. Schelling is not glibly inverting the obvious regularity and clear patterns of the universe in some infantile plea that we should go wild and embrace the Dionysian as an alternative to what we suffer as the iron cage of ordered life. The wild is at the dark ground of the tame, or as Schelling posed the question in *Von der Weltseele*: "How can Nature in its blind lawfulness lay claim to the appearance of freedom, and alternately, in appearing to be free, how can it obey a blind lawfulness?" (I/6, ix). The wild is the free or sovereign progress of the necessary, the creative life of the world. It is the self-organizing, self-unfolding, self-originating, middle voice of nature.

The wild is not, for instance, what Kant in his *Über Pädogik* characterized as *die Wildheit*, wildness or savagery, as "independence from laws."[4] Without discipline and tutelage, humans are unruly. Without dismissing the value of education, it does not follow that the default state, while untutored, is therefore without form and shape and law. The unruly as such is found nowhere—it is a fantasy about what tutelage supposedly conquers (and hence Kant, typical of his epoch, imagined with trepidation whole races of untutored savages). Schelling himself may not always have escaped his culture's prejudices,[5] but his intuition of the autogenesis of being as a system of freedom provides extraordinary resources to rethink what we could mean by the wild, not as the antithesis of the cultured or domesticated, but *the wild as seen from the wild itself* in its natality and self-originating creativity.

At the heart of Schelling's practice of the wild is, to use Deleuze's felicitous phrase, an imageless image of thought, an ontological schema, the intellectual intuition of what belongs by right to the enterprise of thinking. And at the heart of this imageless image of thought is the problem of the imagination as such. This is the central concern of the present study, both as an issue for thinking and as an issue for life practice.

Plan of the Book

Like my first experiment with Schelling's thought, *The Conspiracy of Life: Meditations on Schelling and His Time*, this book does not have to be read in the exact order that it was written. Its three sections can be read as having some relative autonomy in their own right, although all three sections are also variations on the theme of the book (Schelling's practice of the wild as the relationship of time to the imagination).

The first third of the book takes up the problem of time, first in terms of Schelling's *Naturphilosophie*, and then in terms of the vexing problem of philosophy's relation to religion.

In the opening chapter, Extinction, I attempt to frame the book's central problematic (Schelling's "practice of the wild") around the discovery of the problem of catastrophic species extinction in the early nineteenth century. In so doing, I connect the early Schelling's *Naturphilosophie* to the middle Schelling's works by centering on Schelling's emerging theory of temporality. Schelling, along with scientists like Cuvier, was among the first to recognize the implications of the newly emerging fossil record (an event that would have as unsettling an effect as would Darwin's theory of evolution in the later part of the century). The problem of ruination, including the emerging problem of the culture of the industrial human in precipitating the Sixth Great Extinction event, opens a space to appreciate the radicality of Schelling's ethical and ontological efforts to respond to the positivism and anthropocentrism at the heart of the ecological crisis.

The second chapter, Solitude of God, is a close examination of the necessary relationship of revelation to philosophy. Contrary to the received wisdom, which holds that revelation is a religious phenomenon that can be optionally subjected to philosophical scrutiny, Schelling argues that a radical account of revelation is at the heart of philosophy, marking its very possibility. I develop this theme throughout his middle period and conclude with a consideration of his final period, the Berlin lectures on Mythology and Revelation.

The middle part of the book brings Schelling's thought into dialogue with Gilles Deleuze. Those for whom the latter asks too much of them can proceed to the third and final part of the book.

The third chapter concerns Schelling's "image of thought," a phrase I borrow from the famous third chapter of Gilles Deleuze's 1968 *Difference and Repetition*. It names our intuition of what would constitute our sense of what belongs by right (and hence would be relevant to) any possible philosophical enterprise. (It is not itself a concept but rather our intuition of the horizon within which concepts would count or matter to a particular enterprise.) What gives us our sense of what themes, concepts, and arguments belong to a particular philosophical enterprise and which do not? Schelling called this preconceptual sense an intellectual intuition. I bring Schelling into dialogue with Deleuze around the problem of what gives rise to philosophy as such.

The fourth chapter (and the second of two dialogues between Schelling and Deleuze) is about stupidity, one of the problems whose almost utter absence from the Western philosophy's account of itself tells us something disappointing about large parts of that history. It also allows us to appreciate something quite prescient at the heart of Schelling's radical practice of this tradition, including his defense and creative redeployment of some of its subterranean possibilities (Jakob Böhme, Johann Georg Hamann, the Rhineland mystics, the Swabian Pietists, the Kabbala, etc.). In *Difference and Repetition*, Gilles Deleuze asks us to take up the problem of stupidity as a properly transcendental question (indeed a more profound transcendental problem than the one that inaugurates Kantian critical philosophy): "How is *bêtise* (and not error) possible?" I propose to make a contribution to this challenge by examining Schelling on the transcendental problem of the stupidity at the heart of the dogmatic cast of mind. It is important to say from the get-go that by stupidity, neither Schelling nor Deleuze means a lack of intelligence, but rather a fundamental inability to raise questions at an ontological level and to pose problems with sufficient depth. In this chapter I also bring Schelling into dialogue with two of the great thinkers of this problem, Gustave Flaubert and Robert Musil, as well as with the late work of Derrida (especially *The Beast and the Sovereign*).

The final third of the book takes up the problem of Schelling's *Kunstphilosophie* as a way to explore the problem of the imagination in Schelling's thinking and, not incidentally, its challenge to philosophy as such.

The fifth chapter, Plasticity, is centered on Schelling's so-called Munich Address (1807), *Über das Verhältnis der bildenden Künste zu der Natur*, in which Schelling works out the relationship between nature (and by implication *Naturphilosophie* and what we might now call ecological systems theory) and artistic production. This was a problem already at the heart of Schelling's 1800 *System of Transcendental Idealism* and his Würzburg lecture course on the *Philosophy of Art*, but in this dramatic oration, Schelling articulates this relationship with exceptional clarity and insight. In what way are art and nature connected? For Schelling the connective tissue is the *productive imagination* (the productive creativity or coming into being of form).

The final chapter, Life of Imagination, is the key chapter of the whole book. It attempts to undertake the problematic of this entire book (the relationships among science, philosophy, art, history, and religion) as at stake in the problem of the imagination and an overcoming of Platonism's sensible/intelligible diremption. It also attempts to think this problematic in relationship to the turn to positive philosophy that begins with his ancillary work to the great and unfinished *Weltalter* project, namely *On the Deities of Samothrace*. I argue with regard to these texts that it is the *temporality of the productive imagination*, that is, the immense and immensely creative autogenesis of nature, that is at the heart of freedom as the eternal beginning of history (as opposed to Hegel's dialectical progress of history). Special attention is paid to the nature of the image in relation to the imagination. What is the *Bild* in the *Einbildungskraft*?

The book concludes with two special appendices, the first being the first ever publication of a translation of the remarkable 1812 *Letter to Eschenmayer* in which Schelling responds to the latter's critique of the *Freedom* essay. Schelling's exquisite and illuminating response essentially constitutes the first important commentary on the *Freedom* essay. The translation was undertaken with my good friend and colleague, Dr. Chris Lauer, whose own comments on the letter comprise the second appendix.

Part I

Time

1

Extinction

> But this circulation goes in all directions at once, in all the directions of all the space-times opened by presence to presence: all things, all beings, all entities, everything past and future, alive, dead, inanimate, stones, plants, nails, gods—and "humans," that is, those who expose sharing and circulation as such by saying "we," by saying we to themselves in all possible senses of that expression, and by saying we for the totality of all being.
>
> —Jean-Luc Nancy[1]

One Great Ruin

In Schelling's dialogue *Clara* (c. 1810), the doctor advises Clara that if one wants to witness ruins, one does not need to travel to the deserts of Persia or India because "the whole earth is one great ruin, where animals live as ghosts and humans as spirits and where many hidden powers and treasures are locked away, as if by an invisible strength or by a magician's spell."[2] The whole earth is haunted by a mathematically sublime preponderance of ghosts and spirits. Scientists now identify at least twenty mass extinction events, five of which are considered so cataclysmic that they are referred to collectively as the "Big Five" and during which time the conditions for life were cataclysmically altered. Although it is still a matter of some debate, the acceleration of global temperatures and the burgeoning climate emergency due to the increasingly industrial character of human life, the widespread destruction of nonhuman habitats, the alarming rate of rain forest devastation, the unchecked population explosion, and the general degradation of the earth and its resources, is precipitating a sixth. Indeed, the very character of life, given its ruinous history, leaves the earth scarred with fossilized vestiges of former ages of the world, a natural history of the wreckage of past life.

Although Schelling could not have been aware of this current reading of the exuberantly profligate fossil record, nature's luxurious infidelity to its guests was not lost on him. As he mused in *The Ages of the World*: "If we take into consideration the many terrible things in nature and the spiritual world and the great many other things that a benevolent hand seems to cover up from us, then we could not doubt that the Godhead sits enthroned over a world of terrors. And God, in accordance with what is concealed in and by God, could be called the awful and the terrible, not in a derivative fashion, but in their original sense" (I/8, 268).[3] There is something awful and terrible concealed within nature, and it haunts us through its ghostly and spectral remnants. Or to articulate it more precisely: what is haunting about the prodigal ruin of nature is not only that its remnants indicate what once was but is no longer, nor is it enough to say, as does the skeleton at the base of Masaccio's Trinitarian crucifixion (c. 1427) in Santa Maria Novella in Florence: "I once was what you are now and what I am you shall be."[4] It is certainly true that the presence of the vestiges of past life testifies both to the past and to the past's capacity to speak to the future. Both of these moments, however, more fundamentally indicate something awful coming to presence concealed in each and every coming to presence, something awful in which all nature partakes as the paradoxical solitude of its coming to presence.

What all of nature shares, this awful and terrible concealment, is not a common and discernible essence, an underlying substance, or any other kind of universally distributed metaphysical property. Rather, it shares the paradox of coming to presence: each and every coming to presence, what each being shares in its own way, is therefore a solitary coming to presence. Each being is exposed as singular, or, as Schelling adapts Leibniz's *Monadology*, as a monad in the sense of a "unity" or an "idea." "What we have here designated as unities is the same as what others have understood by *idea* or *monad*, although the true meanings of these concepts have long since been lost" (I/2, 64). The monad is the very figure of shared solitude, sharing the awful secret of the absolute as *natura naturans*, yet each in its unique fashion, each singularly. The monad is a particular that is not the instantiation of a higher generality, but rather each monad "is a particular that is as such absolute" (I/2, 64). The community that is nature, a terrible belonging together, is the strange *one*—in no way to be construed as one thing or being—expressing itself as the irreducibly singular proliferation of the *many*, much in the way that Jean-Luc Nancy claims that the "world has no other origin than this singular multiplicity of origins" (BSP, 9).

This is not merely to mark the awful and terrible secret as a limit, as a threshold beyond which thinking dare not pass. As Nancy further reflects, "its negativity is neither that of an abyss, nor of the forbidden, nor of the veiled or the concealed, nor of the secret, nor that of the unpresentable" (BSP, 12). Merely to designate it as such is to designate it exclusively as the "capitalized Other," which marks it as "the exalted and overexalted mode of the propriety of what is proper," relegated to the "*punctum aeternum* outside the world" (BSP, 13).[5] This is precisely what is denied in marking the terrible secret as *the terrible secret of nature*. It is everywhere and therefore everywhere different, the immense dynamic differentiation of the community of solitude that is nature. The "world of terrors" does not merely mark abstractly a limit to thinking. It is the awful secret that expresses itself ceaselessly in and as every single manifestation of the play of nature but which, itself, has no independent standing. It is, as the great Kamakura period Zen Master Dōgen liked to articulate it, the presentation of "the whole great earth without an inch of soil left out."

The plurality of the origin is not only the shared solitude of birth, but also the shared solitude of ruin. This chapter takes as its prompt the phenomenon of mass extinction events, especially the seeming likelihood of the "sixth," but it does so in order to engage in a sustained reflection on Schelling's conception of *natural history*. I argue that for Schelling all history is ultimately natural history, that is, all nature is radically historical and expressive of what Schelling called the unprethinkability (*Unvordenklichkeit*) of its temporality. This consequently subverts the common bifurcation of history into human (or cultural) history and natural (or nonhuman) history. The ascendancy of the Anthropocene Age[6] is widely but erroneously celebrated as the triumph of culture over nature. In order to subvert this duality, I consider carefully the difficult and prescient character of my two key terms: "nature" and "history." For Schelling, the two terms are ultimately inseparable (that is, they belong together as a unity of antipodes). Already established in the early works of his *Naturphilosophie*, and dramatically developed in the 1809 *Freedom* essay and the various drafts of *die Weltalter*, nature is not a grand object, subsisting through time, and leaving behind it the residue of its past. Such a conception characterizes modern philosophy's nature-cide by denigrating nature into an object that can to some extent be pried open from the vantage point of the subjective position of scientific inquiry. This assumes that nature stands *before* us as a vast conglomeration of objects and the eternally recursive laws that govern their manners of relation to each other. As Merleau-Ponty later observed, Schelling "places us not in front of, but rather in the middle of the absolute."[7] Schelling's retrieval of the question of nature is simultaneously the retrieval of its all-encompassing temporality, including its cataclysmic dimensions, but also of its transformative dimension for human thinking.

After a preliminary and orienting reflection on the destructive element of time, I turn to Schelling not by broadly canvassing the vast territory of his thinking, but rather by concentrating on a small number of texts. Although I maintain an eye toward the explosive works of Schelling's transitional period, including the celebrated 1809 *Freedom* essay and the third draft of *The Ages of the World* (1815), I also consider two early writings: "Is a Philosophy of History Possible?" (1798) as well as the beginning of the introduction to *Ideas for a Philosophy of Nature* (1797, revised edition, 1803).

It is in the latter work that Schelling makes the remarkable claim that "Philosophy is nothing but a natural teaching of our spirit [*eine Naturlehre unseres Geistes*]" and that, as such, philosophy now "becomes genetic, that is, it allows the whole necessary succession of representations to, so to speak, emerge and pass before our eyes" (I/2, 39). In moving from the being of our representations to their becoming, to the dynamism of their coming to presence, we become present to the coming to presence of nature itself. That is to say, the organic, nonmechanistic, genetic temporality of nature's coming to presence, that is, nature naturing (*natura naturans*), comes to the fore. This in part yields a mode of access to the vast, paradoxically discontinuous yet progressive history of nature. As Schelling articulates it in *The Ages of the World*: "Therefore that force of the beginning posited in the expressible and exterior is the primordial seed of visible nature, out of which nature was unfolded in the succession of ages. Nature is an abyss of the past. This is what is oldest in nature, the deepest of what remains if everything accidental and everything that has become is removed" (I/8, 244).

Amid the current, heartbreaking, and agonizing explosion of ruin, an event that, while staggering, is hardly unprecedented, accreting the already enormous record of wreckage, I turn to what is "oldest in nature." At the conclusion of the *Freedom* essay, Schelling designates nature as an "older revelation than any scriptural one," claiming that now is not the time "to reawaken old conflicts" but rather to "seek that which lies beyond and above all conflicts" (I/7, 416). Now is not the time to reawaken the "sectarian spirit" (I/7, 335), to pit one position against another, but rather to allow to come to presence, to let reveal itself, that which haunts every possible position. It is time for the most ancient revelation, what is oldest in nature, to come forth. That the sixth biotic crisis is not unprecedented does not make it any less of a crisis. It remains not merely an acceleration of death, but more fundamentally a murderous rampage against nature's natality and hence against its biodiversity (the death of a species is the death of its mode of birth). What does what is oldest in nature enable us to see regarding what is currently the anguish of our earth?

Sifting Through Ruins

If one were to drive one's automobile to a museum of natural history, one could become aware of two ways in which nature is coming to presence. Parking the car, one could enter the museum and, if it were at all comprehensive, one would encounter primordial indications of the ruinous discontinuity of nature in the remnants of earlier ages of the world. Nowhere is this more dramatically evident then when contemplating the fossilized remains of the great reptiles. Their size and power are haunting relics from a scarcely imaginable age, ghosts that speak not only of themselves, but of a lost world, a vastly different ecology of life. Although such things remain issues of scientific debate, the decline of the dinosaurs and the rise of the mammals are generally attributed to the fifth great biotic crisis, occurring some 65 million years ago, perhaps as the result of a collision with a meteor (or some other sudden incursion from space) or a dramatic increase in volcanic activity. In either scenario, the earth's ecological webs were drastically altered and the rate of speciation of macroscopic life was overwhelmed by its rate of extinction.

It is the macroscopic grandeur of the Cretaceous-Tertiary extinction event, with the disturbing and compelling specters of the rapacious Tyrannosaurus Rex, the enormous brontosaurus, and myriad other sublime creatures, that make the paleontology divisions of natural history museums the most gripping and unsettling of haunted houses. The imagination reels all the more when it considers that the magnitude of this loss was greatly exceeded by the third great biotic crisis, which concluded the Permian age. This was the so-called Great Dying, which preceded the fall of the great reptiles by some 180 million years. If one then muses at the spectral record of the species that have died since the last great biotic crisis, the "fearful symmetry" of creatures like the saber tooth tiger or the woolly mammoth, or if one considers the plight of the beleaguered Florida panther or the Himalayan snow leopard, their endangered grasps on life symbolic of the immense pressure on so many different creatures, known and unknown, one then grasps the awful truth of what Georges Bataille meant when he claimed that death "constantly leaves the necessary room for the coming of the newborn, and we are wrong to curse *the one without whom we would not exist.*"[8] Death makes space for the progression of life, the awful secret of what is oldest in nature, haunting nature, progressing anew as *natura naturans*.

When one then, unsettled, leaves the museum of natural history, that repository "where animals live as ghosts and humans as spirits and where many hidden powers and treasures are locked away," and gets back into one's car, and navigates back into a great sea of automobiles, one could reflect that the car's capacity for movement depends on fossil fuels. It consumes the very wreckage that had just been haunting one. Not only that, it partakes in a vast network of human industrial life that is exercising immense, even cataclysmic, pressure on biotic communities. Although the debates continue, there is a growing consensus among biologists that we are amid the sixth great extinction event, with predictions running as high as the net loss of half of all macroscopic species by the end of the century. However, one's automobile is not analogous to a comet or a volcano and catastrophic climate change cannot be attributed to a cosmic or geological accident. We "are" the automobile and the wreckage of the earth is a symptom of our acquisitive wrath. *We are the natural disaster.*

Richard Leakey and Roger Lewin have bemoaned that we "suck our sustenance from the rest of nature in a way never before seen in the world, reducing its bounty as ours grows."[9] Not only is the rise of the human the diminishment of the earth, but the more we diminish the earth, for example, by clear-cutting rain forests for arable land, the more we increase our numbers, which means the more we need to diminish the earth, and so it continues in a deadly progression of self-destructive self-assertion. "Dominant as no other species has been in the history of life on Earth, Homo sapiens is in the throes of causing a major biological crisis, a mass extinction, the sixth such event to have occurred in the past half billion years" (SE, 245).

Is this self, exorbitantly sucking our "sustenance from the rest of nature," that is, the self as the occasion of explosive ruin, clearly distinct from the allegedly dispassionate and inquiring self, which gazes "objectively" at the extravagant expenditure of life that haunts biotic relics? One might be tempted to say that in the first instance one finds oneself *against* nature and in the second instance awestruck, gazing *at* nature. Yet to stare *at* nature, as if it were simply *before* one, is to be no less opposed to nature than to be straightforwardly *against* it. In both instances one finds oneself *in* nature, that is to say, surrounded by nature, amid nature, as if nature were an environment. Nature appears as one's environs when one measures one's relationship to nature as an object distinct from oneself as a subject. Even when one is *in* nature, one is more fundamentally opposed to it, cut off from it, which is the condition of possibility for either gazing *at* it or acting ruinously *against* it.

It is in this spirit of separation, of the Fall from the Garden of Nature, so to speak, that the human gazes even at its own animality as something strange, distant, perhaps lost, and therefore looks at its community with nonhuman animals as somehow beneath the dignity of Aristotle's "animal having λόγος." The human knows itself as the ἀρχή of thinking only as the consequence of having risen *out of* or *above* nature. Such extrication or elevation, however, is not a clean escape, but a fundamental denial or obscuration of oneself, a turning away from oneself in order to elevate oneself above oneself. The vestigial self, left behind in the self's movement toward self-presence, toward the pretense of autonomy, is not yet separated from nature. The "self" that one abandons in order to become distinctively and autonomously a self was not a self-standing self, extricating itself from nature in the act of cognizing itself. It was therefore a "self" at the depths not only of itself, but also of nature, something like what the Zen tradition has called the "original face": "Before your parents were born, what was your original face?" This question, still studied in Rinzai kōan practice (*dokusan*), seeks to initiate the deconstruction of a self that has come to know itself as a self wholly in possession of itself.

The self-possessed subject, the self present to itself, has taken flight from the great life of nature. In the *Freedom* essay, this life on the periphery is characterized, from the perspective of nature, as a sickness, and from the perspective of human life, as radical evil, the original sin of human self-consciousness. This is the paradox of *Selbstbewußtsein* in post-Kantian thought: as soon as the self takes possession of itself, that is, as soon as it identifies with a phenomenal representation (*Vorstellung*) of itself, it loses itself. In direct contrast to the Cartesian position, the subjective self cannot take possession of itself as an object. Self-presentation leaves a trail of relics (a record of presentations), without ever revealing what is being presented.

How then does one think this ghostly subject haunting the relics of oneself? In some strange sense one can designate it as the ground of oneself, but then again, that is also to make this ground objective, to hypostasize it, even in calling it a thing in itself or some object = *x*. It is not a thing, either of the noumenal or the phenomenal sort.

In *The System of Transcendental Idealism* (1800), Schelling argues that the pure self, the self haunting the subject position, "is an act that lies outside of all time" and hence the question "if the I is a thing in itself or an appearance [*Erscheinung*]" is "intrinsically absurd" because "it is not a thing in any way, neither a thing in itself nor an appearance" (I/3, 375). It is literally *unbedingt*, that which has not and cannot become any kind of thing within each thing. Translating *das Unbedingte* as "absolute" always risks hypostasizing *das Unbedingte* and sequestering it to some remote and transcendent realm or sacred precinct (a sacred dimension of being separate from the profane earth). It also risks making *das Unbedingte* too vague, a night when all cows are black.[10] Schelling was clear about this in his earliest writings. In *On the I as Principle of Philosophy or Concerning* das Unbedingte *in Human Knowing* (1795), published when Schelling was twenty years old, he considered this an "exquisite" German word that "contains the entire treasure of philosophical truth." "*Bedingen* [to condition] names the operation by which something becomes a *Ding* [thing], *bedingt* [conditioned], that which is *made* into a thing, which at the same time illuminates that *nothing through itself* can be posited *as a thing*. An *unbedingtes Ding* is a contradiction. *Unbedingt* is that which in no way can be made into a thing, that in no way can become a thing" (I/1, 66). *Das Unbedingte* comes to presence as things, but without revealing itself as anything. Hence in the 1800 *System*, we can see that from the perspective of the objective, seen among things, this is philosophy resuscitated as genetic, as "eternal becoming" and from the perspective of the subjective, it "appears [*erscheint*]," itself a rather spectral verb, as "infinite producing" (I/3, 376), the free play of nature.

The pure self, the original face before your parents were born, is a nonsubject haunting the subject (because the subject is in itself absolutely nothing). It "is" an *Ungrund*, to use Schelling's adaptation of Böhme's phrase (I/7, 407), spectrally present, that is, present in its absence, within *Grund*. How does one face this all-consuming fire within oneself, and within all things, when it emerges, as did Krishna to the despondent Arjuna in the *Bhagavad-Gita*, as Vishnu, the mystery of mysteries, the royal secret, and finally as great time, *mahā kāla*, the world destroyer? In the *Freedom* essay, Schelling argues that *Angst* before the great matter of life, before what is oldest in nature, drives us from the center (I/7, 382). As Schelling developed this thought in *The Ages of the World*: "Most people turn away from what is concealed within themselves just as they turn away from the depths of the great life and shy away from the glance into the abysses of that past which are still in one just as much as the present" (I/8, 207–208).

The face of modern philosophy came to presence in abdicating its original face. In a sense one might hazard to say that its very existence was its original sin (the self falling from nature in its flight toward itself by identifying itself, as we see in chapter 6, with an *image* or *Bild* of itself). Schelling was a close yet worried reader of modern philosophy and a defender of a radicalized Spinoza. One might even say, using the designation carefully, he was a kind of "postmodern Spinoza" in the sense of a Spinoza that has been extricated from the limitations of what was thinkable within modern philosophy's *image of thought*, that is, its estimation of its grounding possibilities and its intuition of what belongs to the modern philosophical enterprise by right.[11] Schelling's Spinoza, unleashing the spectral force of *natura naturans*, with its implicit post-Enlightenment reconfiguration of the philosophy of science, certainly invited ridicule from the prevailing theological orthodoxy, but it also put him at odds with the enlightened scientific standpoint, with its commitment to an autonomous subject (one is tempted even to say, the liberal capitalist subject) as both researcher of nature and moral agent.

In the *Freedom* essay, Schelling provocatively characterized all of modern philosophy as the impossibility of the question of nature even emerging as a serious question. "The entirety of modern European philosophy has, since its inception (in Descartes) the shared deficiency that nature is not present to it and that it lacks nature's living ground" (I/7, 356–357). In resuscitating Spinoza, Schelling breathes life into his thinking, endeavoring to divorce the ground of nature in Spinoza's thinking from any remnant of dogmatism. It is the very freedom at the heart of things that, as *unbedingt*, is what is oldest in nature, always older than its ceaseless coming to presence. It is a life beyond the life and death of things. It expresses itself[12] pluralistically as the shared or "natural" monadic solitude of the life and death of all things. Not only does this deconstruct a Newtonian mechanistic universe, that is, a universe of sheer necessity, adhering to *a priori* laws of nature, but it also deconstructs the autonomous moral subject, adhering to the laws of freedom or to the divine command of a transcendent creator God. "The moralist desires to see nature not living, but dead, so that he may be able to tread upon it with his feet" (I/7, 17).

Time and Nature

Long before he became president, Thomas Jefferson knew of the remarkable mastodon fossils from what was called the Big Bone Lick (formerly in Virginia, now in Kentucky).[13] Yet when he sent Lewis and Clark on their famous trip to the Pacific Northwest, he expected them to find living mastodons. "Such is the economy of nature that no instance can be produced, of her having permitted any one race of her animals to become extinct; of her having formed any link in her great work so weak as to be broken."[14] The laws of nature, whether or not God is their legislator, form a closed and recursive system. All surprises and happenstance are illusions because they are really just manifestations of our ignorance and our lack of mastery of the fundamental rules governing the movement of nature. Catastrophic ruin suggests that God does not know what He is doing or at least that the rules of nature are not ironclad. It has been the Western disposition to err on the side of an omniscient and omnibenevolent God or to have an equally optimistic faith in reductionist materialism.

Georges Cuvier, the French naturalist, did not buy any of this and in 1812 concluded that the fossil record not only indicated that some species were "lost," but that the rate of loss was not exclusively a matter of what is now called *background extinction* (it is the way of species to become *espèces perdues*). He discovered evidence of cataclysmic loss, like the floods that drove Noah to his Ark.[15] Even Darwin, who unsettled the traditional account that held that divine final causality was incompatible with an account of speciation that included, even to a limited extent, the play of chance, opposed the idea that chance was capable of such monstrous profligacy.[16]

Nonetheless, the remnants of sublime squandering, as well as the role of chance mutation in species survival, cast doubt on the unbroken recursivity ("mechanism") of nature. As Iain Hamilton Grant has argued, "the reassuring certainty of a mechanical eternity is removed by the fossil remains of vanished creatures" (PON, 53). Is this not freedom manifesting as contingency?

Schelling's efforts to think through the problem of the relationship between freedom and necessity in nature had to move first through the Kantian critical project. In the *Kritik der reinen Vernunft*, the very appearance of nature excludes the possibility that freedom operates in any way in nature. Rather, the laws of nature manifest the forms of intuition that gather the manifold into experience. We are the legislators of nature. "The intellect [*Verstand*] is itself the legislation [*Gesetzgebung*] of nature, that is, without the intellect there simply would not be any nature, that is, the synthetic unity of the manifold according to rules."[17]

The impossibility of an experience of freedom, of freedom in nature, governs the famous third antinomy, the antinomy of pure reason in its third conflict of transcendental ideas. The antithesis holds, given the very conditions for the possibility of experience, that "there is no freedom but rather everything in the world happens merely according to the laws of nature." If this were the last word, there could be no ethical autonomy and no capacity to obey the moral law. Hence the thesis holds that "causality according to the laws of nature is not the only cause from which the appearances of the world could be collectively derived."[18] The antinomy clearly disallows recourse to an experience of freedom. Nature, in order to appear to reason at all, follows the rules legislated by the intellect. Everything appears as it must appear, in accordance with the rules that gather the manifold into appearance. Freedom, the sine qua non for practical philosophy, does not and cannot appear in nature. Instead, it commands from the noumenal kingdom of ends, from beyond the rule of nature. This is Kant's famous bifurcation: human life simultaneously dwelling under the laws of nature and the deontological citizenship in the autonomous kingdom of ends.

With this account, it makes little sense to speak of natural history as anything other than the chronology of rule-bound happenings. There can be no real sense of history in the sense of what happened. What happened did not happen by happenstance. It had to happen. But the fossil record!

The possibility of anything like a living natural history (rather than the mechanical natural chronology) opens up in the third *Kritik*, which Schelling regarded as "Kant's deepest work, which, if he could have begun with it in the way that he finished with it, would have probably given his whole philosophy another direction."[19] Its project is nothing less than to try to reflect from the space that opens up in the "incalculable chasm [*die unübersehbare Kluft*]" between nature and freedom, the laws of appearance and the nonappearance of freedom. In aesthetic judgment (that is, in the exercise of taste), one reflects on the pleasure that one senses operating at the ground of and at odds with the laws of nature, its "reference to the *free lawfulness* of the imagination [*die* freie Gesetzmäßigkeit *der Einbildungskraft*]."[20] For example, Kant objects to William Marsden's claim in his *History of Sumatra* that when, amid the overwhelming prodigality of "free beauties" in the Sumatran forests, he discovered the beauty of a nice and tidy pepper patch, he had found real beauty, as opposed to the chaos of the jungle. For Kant, what made Marsden judge the pepper patch as beautiful was that it was unexpected, a surprise in the jungle. If he were to gaze exclusively at the pepper patch, he would soon grow bored, as the free play of nature that unexpectedly came to presence became the rule that pepper patches are the only beautiful things in the Sumatran jungle. Soon Marsden's attention would wander and return to the "luxuriance of prodigal nature, which is not subjected to the coercions of any artificial rules" (KU, §22, 86). When one finds oneself taking pleasure in the aqueous undulations of a waterfall, or the dancing flames of a campfire, or the quietly dynamic flow of a babbling brook, does one not delight in one's incapacity to comprehend the principle at play in their intuition

(KU, §22, 85–86)? It is the unbidden pleasure taken in the free play of nature's rule.

It is as if nature presented itself in the element of water, capable of taking any form, but having no form of its own. In the water consciousness of aesthetic judgment, one does not seek to explain nature, but rather one becomes aware of the wonder of nature, of the miracle of its coming to presence. In the sublime this dynamic is intensified as the immeasurability (*Unangemessenheit*) and boundlessness (*Unbegrenztheit*) of this freedom, "makes the mind tremble," filling one with a feeling of astonishment, respect, the shudder of the holy, and a quickening of life. One could say that the possibility of a living natural history, itself only possible with the shattering of the paradigm that dictates that nature's temporality is recursive, like a clock, begins to suggest itself in the dawning of the sense of a whole that holds together the antipodes of nature and freedom.

For Kant, however, this whole does not come entirely to the fore. Freedom is but the feeling of the moral law within projected on the starry heavens above. He does not yet know Alyosha Karamazov's more difficult joy: "The silence of the earth seemed to merge with the silence of the heavens, the mystery of the earth touched the mystery of the stars. . . . Alyosha stood gazing and suddenly, as if he had been cut down, threw himself to the earth. He did not know why he was embracing it, he did not try to understand why he longed so irresistibly to kiss it, to kiss all of it, but he was kissing it, weeping, sobbing, and watering it with his tears, and he vowed ecstatically to love it, to love it unto ages of ages."[21] For this to happen, Kant needed to remember his original face and therefore that the pleasure and shudders of nature are not found within ourselves but within the original face that always already shares and touches the original face of nature.

For Schelling, operating in the unity of the "incalculable chasm" articulated by Kant, natural history speaks from nature as the progression of freedom and necessity. As Grant articulates it: "Natural history, then, does not consist solely in empirical accounts of the development of organizations on the earth's surface, nor in any synchronic cataloguing of these. Its philosophical foundations make it a science that attempts to straddle the gulf between history, as the product of freedom, and nature, as the product of necessity" (PON, 18).

In a small essay from 1798, "Is a Philosophy of History Possible?", Schelling defines history according to its etymology as "knowledge of what happened" (I/1, 466). Hence, "what is *a priori* calculable, what happens according to necessary laws, is not an object of history; and vice versa, what is an object of history must not be calculated *a priori*" (I/1, 467). History in relationship to nature, then, is the play or chance occurrence of freedom in nature, not in the sense that nature becomes a chaotic free for all and collapses into mere "unruliness [*Gesetzlosigkeit*]," but rather that it is not merely subject to rules. There is also variance and deviation (*Abweichung*) (I/1, 469), instances where the rule did not hold. In this sense, "history overall only exists where an ideal and where infinitely-manifold deviations from the ideal take place in individuals, which nonetheless remain congruent with the ideal as a whole" (I/1, 469). In other words, progression assumes deviation from the ideal and, as such, is the expression of the course of a free activity that cannot be determined *a priori*. "What is not progressive is not an object of history" (I/1, 470).

Schelling, perhaps not to his credit, did not regard animals as having history "because each particular individual consummately expressed the concept of its species" and there was therefore no "overstepping of its boundaries" and no "further construction on the foundation of earlier" individuals (I/1, 471). An individual bear acts in accordance with how bears as such generally behave. Animals do not need to act freely because they are not subject to the original sin of self-consciousness—an animal does not require therapy to relax the pernicious hold of Lacan's mirror stage. Since an animal is not tempted to identify self-consciously with its *imago*, it does not experience its freedom in overstepping its *imago*. What is freedom from the perspective of *natura naturans* is tacit necessity from the experience of animals because they are not self-consciously free. That being said, the nonhuman animal community is full of surprises, and it is wise not to speak too confidently about them. Nonetheless, Schelling's sensitivity to the general problem, beyond the complex and vexing problem with nonhuman animals, remains acute: "where there is mechanism, there is no history, and vice versa, where there is history, there is no mechanism" (I/1, 471).

Even if one unwisely granted that animals do not have a history, that is to say, a history that belongs to animals as animals, *it does not follow that nature does not have a history of animals*. Even if animals do not have a specifically historical consciousness, there is a natural history of animals, just as there is a natural history of everything, from subatomic particles to black holes. Just as deviation from oneself is the possibility of humans having history, that is, of acting freely, nature's progressivity is its capacity not to be held hostage to its manifold appearances, to deviate from itself ever anew. As Schelling articulated in his introduction to *Ideas for a Philosophy of Nature*, written around the same time as the small piece on history, "Nature should be visible spirit, spirit should be invisible nature" (I/2, 56).

The very idea of *Naturphilosophie* is not to define nature as a philosophical object, but rather to recover nature as "the *infinite* subject, i.e. the subject which can *never* stop being a subject, can never be lost in the object, become mere object, as it does for Spinoza" (HMP, 99/114). As such, its translation as "the philosophy of nature" is potentially misleading. *Naturphilosophie* in Schelling's sense is more like doing philosophy in accordance with nature (not as an elective philosophical topic originating at the whim and command of the *res cogitans*). It is not therefore a kind of philosophy, or a topic within philosophy, but rather a gateway into the originating experience of philosophizing.

In his startling and exceptionally lucid thought experiment at the beginning of his introduction to *Ideas*, we see Schelling orchestrating the "originating [*enstehen*]" of this subject "before the eyes of the reader" (I/2, 11). This origination and "coming to the fore" happens by simply reflecting on the nature of philosophy itself. Is philosophy any particular philosophy? If philosophy is not any particular philosophy, what is it that we do that includes the remnants of all philosophies heretofore, but which is exhausted by none of them? Philosophy, one could say, is the free, historical act of philosophizing, not any particular philosophy. Or: the subject of philosophy is reducible to no particular philosophical objects although it is expressed in all of them. Hence, Schelling claimed in the first (1797) version of this introduction that "the idea of philosophy is merely the result of philosophy itself, but a universally valid philosophy is an inglorious pipedream [*ein ruhmloses Hirngespinst*]" (I/2, 11). One might say that the very desire to make *a* philosophy *the* philosophy is itself undignified illness, recoiling in anxiety from the freedom of philosophizing.

This origination of the subject also comes to the fore when one extends the subject of philosophy to nature. It begins with reflecting on being able to ask simple questions like: What is nature? How is nature possible? If nature were merely mechanistic, one could not ask this question. Reflection by its very nature is what David Wood has called "the step back, the promulgation of negative capability," which resists "unthinking identifications."[22] In reflection we divorce ourselves from nature, separating ourselves from an absorption in nature. It could be the "No!"—the abrupt eruption within language of a fundamental refusal—that separates us from absorption in the present, or it could be the sudden dawning of a living doubt that dissipates the hold of the obvious, or it could be the surprise discovery that Plato's prisoner makes when he turns around and, seeing the heretofore concealed manufacture of the present, realizes that the images of the present are questionable and mysterious, filling him with wonder and arousing the desire for further exploration. All of these are examples of the sudden and epiphanic interruption that breaks the spell of presence. We ask if what we see of nature and what we already think of nature is sufficient to appreciate nature. As if one were in Plato's cave, reflection, the eruption of the radically interrogative mode, "strives to wrench oneself away from the shackles of nature and her provisions" (I/2, 12).

However, one does not tear oneself away from nature as an end in itself. One rejects the grip of nature as a means to grasp more fully nature itself. Mere reflection, that is, reflection for the sake of reflection, is, accordingly and in anticipation of the *Freedom* essay, *eine Geisteskrankheit des Menschen* (I/2, 13). *Eine Geisteskrankheit* is a psychopathology or mental disease, literally, a sickness of the spirit. One pulls away from the center of nature and its stubborn hold and retreats to the periphery of reflection. From the periphery, nature is no longer the subject (the center), but we assume that we are the subject, that we are the center. If one remains on the periphery, separated, alone in the delusion of one's ipseity, this is the experience of sickness and radical evil. In the language of the introduction to the *Ideas*, when reflection reaches "dominion over the whole person," it "kills" her "spiritual life at its root" (I/2, 13). Reflection has no positive value in itself. It only as has a "negative value," enabling the divorce from nature that is our original but always mistaken perspective, but it should endeavor to reunite with that which it first knew only as necessity. Reflection is "merely a necessary evil [*ein notwendiges Übel*]" that, left to itself, attaches to the root, aggressing against the very root of nature that prompted the original divorce from the chains of nature. Philosophy, born of the abdication of nature, is the art of the return to nature. In such a return, the self of nature, so to speak, comes to fore as the eternal beginning of nature. Just as the root of philosophy is exhausted in no exercise of philosophy, the root of nature is exhausted by none of its expressions. The history of nature is the unfolding legacy of what is always already gone in all that originates and comes to presence. Since it expresses everything in its coming to presence, as well as the mortality and emptiness (lack of self-possessed intrinsic being) of all presence, its origination and points of access are as multiple as things themselves.

In the temporality of nature, what is oldest in nature, whose remnants are its history, but which in itself remains always still to come, promises fatality as the truth of natality. "The eternal beginning [*der ewige Anfang*]" begins ever anew because of the generativity of finitude. Bereft of the radical interruption that is both finitude and the incessant natality of the future, however, nature becomes a nightmare realm populated by angels, perfect beings, wholly obedient to their forms. Angels do not partake in history; Klee and Benjamin's angel of history is the murderous face of history that falsely and ruinously imagines that it has become immanent to itself, as it in some way does in Hegel. Already in the 1798 history essay, Schelling dismissed angels as "the most boring beings of all" (I/1, 473) and almost thirty years later said of Hegel's God that "He is the God who only ever does what He has always done, and who therefore cannot create anything new" (HMP, 160/160). One might say, then, that Hegel's great Angel is not merely boring, but rather *the wrath of the boring*. In general, Hegelian philosophy, for all its brilliance and elasticity, cannot address the impetus of positive philosophy: if Hegel thought that there could not be anything outside of the concept, it cannot think the life of the eternal beginning (*ewiger Anfang*). "The whole world lies, so to speak, in the net of the understanding or of reason, but the question is precisely how did it come to be in this net? For there is still manifest [*offenbar*] in the world something other and something *more* than mere reason, even something that strives to go through and beyond these limits [*etwas über diese Schranken Hinausstrebendes*]" (I/10, 143–144). Hegel's dialectic is still part of this net, and is perhaps its angelic guardian.[23] Life resists the very forms that it engenders and that is the mystery of its creativity and vitality.

In the late lectures on the philosophy of revelation, Schelling, in his defense of the spirit world (*Geisterwelt*), dismisses the possibility that there are beings or creatures called angels. "The angels could not be creatures because they are pure potencies or possibilities . . . mere possibilities are not created; only the concrete or the actual are created."[24] Accordingly the representations of angels have not been successful precisely because are not endowed with the temporality that enables them to enter into history. Only the dark angels, which actualize themselves insofar as humans aspire to become the center of gravity, have mythological force and in this way they are real to us.

> [Satan's] demonic nature is an eternal avidity—ἐπιθυμία [appetite, yearning, longing, concupiscence]. The impure spirit, when he is external to humans, is found as if in a desert where he lacks a human being in which to actualize his latent possibilities. He is tormented by a thirst for actuality. He seeks peace but does not find it. His craving [*Sucht*] is first stilled when he finds an entrance into the human will. Outside of the human will, he is cut off from all actuality—he is in the desert, that is, he is in the incapacity [*Unvermögenheit*] to still his burning longing for actuality. (U, 648)

Hence, Milton's Satan is fully alive precisely because this principle emerges in time as an image, but the problem of evil is even more forcefully present, as we see in the final chapter, in characters like Shakespeare's Macbeth or Melville's Ahab, characters that invert the relationship between existence and ground and become themselves the ground and assume for themselves the principle of the eternal beginning. Satan leaves the desert insofar as we desire to rule the universe from the periphery. Satan, fully alive not as a creature, but as a "principle" that actualizes itself in time as the faculty for evil, does not originate in the human fancy as a way of illustrating or exemplifying the principle at the heart of sickness and evil. "Mythology does not originate in the free invention of human arbitrariness, but rather in the inspiration of a real principle" (U, 648). This principle emerges as the very real *time of an image*, in this case, the elemental image or *Urbild* of the desert yearning to express itself by tempting us, so to speak, to lord ourselves over the earth.

The time of nature is its abyssal past returning as the *Unvordenklichkeit* (unprethinkability) of the future. As such, the past is not therefore the record of the continuous history of some grand object = x. It is rather the evidence of an occasionally catastrophic record of discontinuity, and the history of the ruptures of time that persist, for example, in the intersection of different ages (or different economies and ecologies of being). In *The Ages of the World*, Schelling offered the example of comets, those mysterious emissaries from another age:

> We still now see those enigmatic members of the planetary whole, comets, in this state of fiery electrical dissolution. Comets are, as I expressed myself earlier but would now like to say, celestial bodies in becoming and which are still unreconciled. They are, so to speak, living witnesses of that primordial time, since nothing prevents the earlier time from migrating through later time via particular phenomena. Or, conversely, nothing prevents a later time from having emerged earlier in some parts of the universe than in others. In all ages, human feeling has only regarded comets with a shudder as, so to speak, harbingers of the recurrence of a past age, of universal destruction, of the dissolution of things again into chaos. Evidently, the individual center of gravity (the separate life) in a comet is not reconciled with the universal center of gravity. This is demonstrated by the directions and positions of their paths that deviate from those of the settled planets. (1/8, 329–330)

Nature Is Bizarre

Comets are strange, as strange as the mastodons that Jefferson hoped that Lewis and Clark would find in their travels. That they did not find mastodons was no less strange than that they did find grizzly bears. Nancy, in his beautiful way, reflected that "'Nature' is also 'strange [*bizarre*],' and we exist there; we exist *in* it in the mode of a constantly renewed singularity" (BSP, 9). The term *bizarre* is of uncertain origin, perhaps originating from the Basque word for beard, perhaps in so doing recording the strangeness of the appearance of bearded Spanish warriors. In addition to the element of wonder and surprise, it also speaks to the dignity and grandeur of that surprise. This "'strangeness' refers to the fact that each singularity is another access to the world" (BSP, 14).

The sudden reemergence of the dignity of nature's strangeness reunites (without dissolving singularity into identification) thinking with the nature that it had forsaken. In the famous 1797 *System* fragment, written by Hegel, Hölderlin, and Schelling, the dignity of nature's strangeness, something that Hegel would in an important respect later renounce, was designated the practice of *natural religion*, and Schelling, even in his final period (the Berlin lectures on mythology and revelation), never loses sight of it.

What Schelling calls for in this origination of the strangeness of nature, in the cultivation of natural religion, is a *practice of the wild*. I take this term from the North American—or better, Turtle Island (its name in the eyes of the Haudenosaunee or Iroquois Confederacy—poet and essayist Gary Snyder. In the final chapter called "Grace" in his duly celebrated *Practice of the Wild*, he explains that at his house they say a Buddhist grace, which begins, "We venerate the Three Treasures [teachers, the wild, and friends]."[25] The three treasures are universally acknowledged by all negotiators of the Buddha Dharma to be the Buddha, which Snyder, using his own *upāya* or skillful means, renders as "teachers," the Sangha, the community of practitioners, whom Snyder renders as "friends," and finally, and most strikingly, the Dharma, which Snyder renders as "the wild."

In what manner can the Dharma, the very matter that is transmitted from Buddha Dharma to Buddhist negotiator, be translated as *the wild*?

It all depends on how one hears the word "wild."

Typically "wild" and "feral" (*ferus*) are "largely defined in our dictionaries by what—from a human standpoint—it is not. It cannot be seen by this approach for what it *is*" (PW, 9). Hence, a wild animal is an animal that has not been trained to live in our house (undomesticated) and has not been successfully subjected to our rule (unruly). We look at the wild, indeed the nature left to its own movement, as opposed to our mode of dwelling, as the antipode to culture.

But what happens if we "turn it the other way"? What is the wild *to the wild*? What is nature seen from its own center, that is, from itself as its own subject, not viewed from the periphery as we invert it into an object as the center becomes our own now detached and alienated subjectivity? The wild from the perspective of the wild—Schelling's nature from the perspective of its own *living* or wild ground—is no longer surrounding us as a place *in* which are located. We *are* the earth's bioregions. Animals become "free agents, each with its own endowments, living within natural systems" (PW, 9). As Snyder begins to explore this turn, he indicates the ways in which the wild "comes very close to being how the Chinese define the term *Dao*, the *way* of Great Nature: eluding analysis, beyond categories, self-organizing, self-informing, playful, surprising, impermanent, insubstantial, independent, complete, orderly, unmediated . . ." (PW, 10). And Dao, as we know from the rich interpenetration of Mahāyāna and Daoist traditions in East Asia, is "not far from the Buddhist term Dharma with its original senses of forming and firming" (PW, 10). The early Daoists spoke of Dao as "the great mother" and Schelling emphasized that the etymology of *natura* is that which has been born, and that *natura* thereby speaks of prodigal natality.

The Dharma and the Dao are the sovereign ungrounding of life, such that as Schelling argued in 1806, "immanence and transcendence are completely and equally empty words [*völlig und gleich leere Worte*]" (I/2, 377). There is no outside to the wild—that is the illusion of the ego position—and hence there is equally no inside to the wild, yet its autogenesis is the wildness that still lingers in art, that calls forth religious care, that inspires the respect that endows scientific study with its careful love of details, that animates the careful researches into the historical past, and that is "the authentic mystery of the philosopher" (I/8, 201). In this sense I speak of philosophy, at least in the sense of the activation and ritualization of this experience of philosophizing without idols, thinking that cannot represent to itself the mystery of its intuition of itself, *as a practice of the wild*. Schelling's practice of the wild is to think and dwell with and on the sovereign autogenesis of nature. And as we develop more fully in the final chapter, "the wild is imagination."[26]

What are we to make of ourselves as the occasion of the sixth great extinction? This question is inseparable from the management of our cities, the cultivation of our bodies, the way that we read books—everything depends on the recuperation of the wildness that we share with all of being, the wildness (beyond the ruinous dichotomy of wild and civilized) that is our abdicated being together with and as nature.

The Image of the Unimaginable

Baron Georges Cuvier (1769–1832) is remembered for insisting that nature progressed not by evolution, as he thought Jean-Baptiste Lamarck (1744–1829) mistakenly argued, but rather by catastrophic interruptions, including the cataclysmic flood that drove Noah to his ark (what he surmised was a real disastrous event concealed within its mythological Biblical account). Cuvier's geology issued from the earth's record of the strata of destroyed forms of life layered on destroyed forms of life, each stratum piled on the ruins of the previous mass extinction event. Comparing the mummified cats that Napoleon and his men had looted from Egypt to contemporary cats, Cuvier detected no significant structural differences and mistakenly took this as evidence that species do not change over time nor do they perish slowly as they are increasingly unable to adapt to what Lamarck had called *l'influence des circonstances*. Species disappear all at once by way of devastation. As Martin Rudwick explains:

> For he seems never to have considered the possibility that a species might become extinct gradually, by slowly losing a long battle to maintain its numbers in a natural habitat. Any species was for him such a well-adapted "animal machine" that only a catastrophic event could make it go extinct: under the impact of any more gradual changes in the environment, a well-adapted species would simply migrate and survive. In fact, for Cuvier, as his texts . . . show repeatedly, species did not just become "extinct" [*éteints*]—though he did sometimes use that word—but, rather, were "destroyed" [*détruits*] or "wiped out" [*anéantis*].[27]

After Darwin we can appreciate that Cuvier's insistent hypthesis that biota were but well-adapted "animal machines," enduring through time and only departing all at once through a ruinous interruption of their living conditions, is not supported by the evidence. Evolutionary theory allows us to appreciate the gradual drifts of speciation. Despite the profundity of his insistence on catastrophic time (the ultimate discontinuity of natural history), Cuvier's mechanical life forms are what Schelling will call dead forms (see chapter 5) because Cuvier is unable to think that life itself is the whole of time (the simultaneity of past, present, and future), that time is the very animating (*beseelende*) force that we call life (see chapter 3). This does not mean that Schelling argued for an independent life-force or *Lebenskraft*—life does not enter form from the outside or even the inside. *Life is the temporal positing of form. Mere form is but the dead hull of life.* Nonetheless, Cuvier's insistence on the revolutionary nature of geological time, namely that the strata of the earth speak to the sudden violence of cataclysmic alteration, is on the cusp of a radical intuition of the problem of the earth. This is hinted at in the remarkable "Preliminary Discourse" to his *Ossemens fossils*:

> When a traveller crosses fertile plains, where the regular course of tranquil rivers sustains abundant vegetation, and where the land—crowded with numerous people and ornate with flourishing villages, rich cities, and superb monuments—is never disturbed unless by the ravages of war or by the oppression of powerful men, he is not tempted to believe that nature has also had its civil wars, and that the surface of the globe has been upset by successive revolutions and various catastrophes. But these ideas change as soon as he seeks to excavate this ground that today is so peaceful, or to climb onto the hills that border the plain. (GC, 186–187)

The earth itself and its strata whose surfaces bespeak "its civil wars" makes for a strange scientific object, as if we could have ever kept the earth at an objective distance while remaining ourselves (subjects having their own being inquiring into the earth as if it were a separable object). Its earthquakes are already the earthquakes within us, the wild at the heart of all images, although it is impossible to imagine Cuvier appreciating John Sallis' tremulous articulation of the earth as "that to which the things that arise, that come forth into the light, are brought back, sheltered, and finally entombed."[28]

It should come as no surprise, then, that Cuvier rejected Schelling's *Naturphilosophie* as illogical.[29] In a sense, Cuvier was right. The question of nature defies what we have traditionally meant by logic (although it does not exclude the possibility of rethinking the enterprise of logic as such). Right from the get-go Schelling demonstrates that beings are not as such by owning their own being and by being identical to themselves.[30] Cuvier concluded that Schelling's tendency to regard nature as a "living organism" was experimentally unverifiable and that such metaphysics could only discredit the natural sciences (GCV, 136). In this sense, Cuvier's allergic reaction to *Naturphilosophie* was nothing unusual, but rather ultimately emblematic of modern science's inclusion within modern philosophy in general and that, as such, "natural" science was oblivious to the *natural.*

Although Cuvier the geologist and fossil specialist was close to the earth (its "civil wars"), he kept its cataclysmic temporality, indeed, its temporality as such, at a safe distance, and from the periphery did not let it speak fully and did not unleash the λόγος in geology. This distance can also be instructively detected in the work of his younger brother, Frédéric Cuvier (1773–1838), the illustrations for whose *De l'histoire naturelle des cétacés* [*On the Natural History of Whales*] (1838) Herman Melville's Ishmael evaluates in *Moby-Dick: Or, the Whale* (1851). Although I return to this work, perhaps the most consummate example of what we might call the *Schellingian novel*, in the concluding chapter, we can already appreciate in these short passages the death of nature in the gaze of the Cuvier brothers, despite their at times startling proximity to it.

It is remarkable to note that Frédéric Cuvier, a zoologist and the chief keeper of the menagerie at Paris' Muséum national d'histoire naturelle, wrote about whales, including the spermaceti, which Ishmael often calls *leviathans* and *monsters*. The latter term is particularly apposite for both Ishmael's verbal painting (imagination of the whale) as well as our present purposes because whales enable us to glimpse in intellectual intuition the monstrosity of the ocean and the earth. This is not because they are aberrations, but rather because their life brings their perception as well as their conception to the limit, bringing us to the brink of what Sallis has called a *monstrology*:

> For what is to be inscribed or traced, the showing of things themselves, is not reducible to presence, does not present itself, does not offer itself to an intuition determined as such by correlation with presence. A discourse that would inscribe the showing of things themselves cannot but transgress the limits of mere explication and violate what was to have been the principle of all principles. The inception of such a discourse marks the passage of phenomenology over into monstrology. (FI, 104)

Moby-Dick, it could be said, is a saying of the monstrous—the monstrosity of whales, the monstrosity of Moby Dick's whiteness, the monstrosity of the sea itself, which allows us to float on it as it shelters its myriad creatures in their prodigal differentia, but which is always ready to swallow us up utterly, at any minute, refusing "resurrections to the beings who have placelessly perished without a grave" (chapter 7, The Chapel).[31] This ocean, beyond the seeming comforts of *terra firma* (Cuvier's "civil wars" already dispel the illusion of the *firma*), brings language and truth to the limits of its discursive competence: "but as in landlessness alone resides the highest truth, shoreless, indefinite as God—so, better is it to perish in that howling infinite, than be ingloriously dashed upon the lee, even if that were safety" (chapter 23, The Lee Shore)! This is not the story of Fr. Mapple's repentant Jonah, who, having returned to the fold of his proper nature, is cast by the whale safely back onto the shore. This is the leviathan of which the whirlwind speaks to Job:

> Can you catch him with a fishhook or hold his tongue down with a rope? Can you put a ring in his nose or pierce his jaw with a barb? Will he entreat you at length?
>
> Will he speak with you softly? Will he agree with you to be your slave forever? Will you play with him as you would with a bird or keep him on a string to amuse your little girls? Will a group of fishermen turn him into a banquet? Will they divide him among the merchants? Can you fill his skin with darts or his head with fish-spears? If you lay your hand on him, you won't forget the fight, and you'll never do it again! (Job 41:1–8, *Complete Jewish Bible*)

Frédéric Cuvier's whales, like John James Audubon's *Birds of America*, in which Audubon killed and stuffed his otherwise flighty birds so that they would stay still so that he could illustrate their forms accurately, are also literally dead. Audubon used fine shot to slay his birds, but then arranged them in a "natural" (!) pose. Cuvier's illustration of the male spermaceti whale depicts an impressively large specimen lying expired on the shore, removed from its monstrous home. (At least there is no pretense of returning it to a more natural pose!) Three men work on its enormous carcass while a fourth tends to his boat, and all of them help give us some sense of the leviathan's scale. The whale's dead mouth has opened, revealing the (male) teeth with which he does battle with giant squids. Cuvier, who occupied a chair of comparative physiology, captures the specific form of the whale with acute detail and precision. Ishmael retorts, however, that "Cuvier's Sperm Whale is not a Sperm Whale, but a squash . . . the great Leviathan is that one creature in the world which must remain unpainted to the last" (chapter 55, Of the Monstrous Pictures of Whales).

The chapter title speaks of such pictures as monstrous, as if they were monstrous representations of the monstrous. But what makes such representations in themselves monstrous? What is the monstrosity at the heart of representation itself—"that one creature in the world which must remain unpainted to the last"? The monstrosity of Cuvier's illustration of the spermaceti is precisely the monstrosity of an utter lack of monstrosity, as if Moby Dick is best understood as physeter macrocephalus, the physeter with the big head. When Ishmael conducts his own research into cetology, it is not without extensive attention to detail, but with the caveat that

> To grope down into the bottom of the sea after them; to have one's hands among the unspeakable foundations, ribs, and very pelvis of the world; this is a fearful thing. What am I that should essay to hook the nose of this leviathan! The awful tauntings in Job might well appall me. "Will he (the leviathan) make a covenant with thee? Behold the hope of him is vain!" (Chapter 32, Cetology)

Representation is already the tacit hope of a covenant, that the form (of the whale) is its being, that its life is a mere accident (its form is its idea, its meaning, making its existence accidental). Morphology, however, remains too peripheral to the κῆτος. The form is what it is, what Frege called its *Sinn*, while its life is its mere existence, that is, again with Frege, its *Bedeutung*, its reference to something in the world. Cuvier illustrates the *Sinn* or sense (idea) of the whale. Ishmael, to the contrary, insists that it is the life of the whale as such, the sheer monstrosity of that fact, that gives rise to thought, although precisely at the limit of the whale's comprehensibility. We study the form of the whale precisely because it is how we approach its awe-inspiring monstrosity, indeed, monstrosity as such. The whale comes as itself and it does not mean anything beside itself, but it emerges in such a way that the oceanic monstrosity of emergence as such shimmers through the whale.

Cuvier's representations are profane simply because they belong to representation as such. This does not mean that science should not assiduously study the look of things, just as Ishmael attempts his cetological studies. It is, rather, that such studies are not reducible to the profane cataloguing of the formal features of biota. It is instead, as Schelling insists, the contemplation of the mystery of life, not the accretion of dead images (the profanity of the reduction of the image to representation):

> Science is the knowledge [*Erkenntniß*] of the laws of the Whole, therefore of the Universal. Religion, however, is the contemplation [*Betrachtung*] of the particular in its being bound up with the All. It ordains the natural scientist [*Naturforscher*] as a priest of nature through the devotion [*Andacht*] with which she cares for the individual details. Religion assigns to the scientist the limits, posited by God, of the drive for the Universal, and thereby mediates science and art through a sacred copula [*heiliges Band*]. Art is the imagination [the presenting as an image, the one expressing itself as a particular image—*Ineinsbildung*] of the universal and the particular. [I/7, 141][32]

The monstrosity at the heart of the scientist's devotion to *things just as they are (in their becoming)* is the same monstrosity out of which the imagination creates, revealing art itself to be our relationship with the oceanic, our participation in the monstrous creativity of *natura naturans*. Moby Dick reveals nothing concrete about the ungrounding ground of beings, but it does reveal the monstrosity of that ungrounding. This is precisely what Moby Dick's whiteness reveals: it does not refer to some idea beyond the fact of its whiteness. Rather the darkness of death shimmers through the shining of the leviathan's white skin: "It cannot well be doubted, that the one visible quality in the aspect of the dead which most appalls the gazer, is the marble pallor lingering there; as if indeed that pallor were as much the badge of consternation in the other world, as of moral trepidation here" (chapter 42, The Whiteness of the Whale).

Death shimmers through the skin of the earth as if it were a concealed ground that is not supporting us, but like the ocean, momentarily tolerating our efforts to stay afloat on it. "Is it that by its indefiniteness it shadows forth the heartless voids and immensities of the universe, and thus stabs us from behind with the thought of annihilation, when beholding the depths of the Milky Way?" (chapter 42). Nature eludes representation—mere idolatry—but is rather *the life of the imagination*, the mysterious productivity of images "so that all deified Nature absolutely paints like the harlot, whose allurements cover nothing but the charnel house within" (chapter 42).

2

Solitude of God

Denn schwer ist zu tragen Das Unglück, aber schwerer das Glück.
—Hölderlin, *Der Rhein*

Philosophical Revelation

My thesis in this chapter is as audacious as it is simple: Schelling's celebrated 1809 *Freedom* essay, despite the clear ontological fabric that Heidegger rightly retrieves and explicates in his various Schelling lectures, is more fundamentally the exposure of humanity's exposure to Being as *revelation*. In a word, it presents revelation to philosophy. It reveals revelation to philosophy as belonging to philosophy.[1] To be clear: this does not first assume the possibility of revelation and then inquire about such a possibility philosophically. Rather, it is the exposition of the character of exposition and, as such exposes exposure as having the character of revelation; it reveals revealing itself at the heart of humanity and therefore also at the heart of philosophizing. The philosophical revelation of revelation does not, however, reveal to human beings a particular set of predicates that characterize humanity because the preeminence of such a demand is precisely what defines humanity as *Angst* before the fiery ground of revelation. Rather, revelation is, even more pointedly, the revelation of the solitude of God as first philosophy.

This claim is in no doubt scandalous not only to Schelling adherents, but even more pointedly to both traditional theological traditions and the generally atheistic disposition of contemporary culture. This is not, however, a reactionary attempt to reinsert God into human affairs precisely at the moment of its twilight. Schelling's philosophical religion is prophetic; nothing like it has ever come about in anything like a significant scale.

Let us be clear: the implication of the *Freedom* essay, already suggested in the aptly titled 1804 *Philosophy and Religion* essay, not only denies to theology its traditional aspiration of subordinating philosophy to theology—Schelling does not begin with faith and then seek understanding—it even more fundamentally denies to theology its even more basic dream of autonomy from philosophy. (To be fair, it also denies to philosophy that it can ever fully escape the problem of religion.) Revelation belongs first to philosophy, indeed, it is first philosophy itself, and at the heart of revelation, naming the very revelatory force of revelation itself, is the solitude of God.

What would it mean to insist on solitude not only as a philosophical virtue, but also as what remains to be said of our theological aspirations? As such, solitude is not exclusively cultivated as a virtue for the philosopher and for the possibility of philosophizing. Rather, it would demand that we dare think the solitude of God, that we find the requisite generosity, and that we still our anxiety before the fiery ground of life itself, to grant God the possibility of solitude. (Note: The mood of the philosophical inquiry that reveals in its innermost progression the solitude of God cannot be a mood of neediness and anxiety. Solitude of any kind cannot be found in the terms set by need or anxiety.)

I note here already that Schelling's phrase "fiery ground" recalls the all-consuming fire of Elohim who, as the burning bush, summoned Moses (Exodus 3:4),[2] that is, the fire that appears in the conflagration of appearance as such. This, Schelling tells us in *The Ages of the World*, is the "doctrine of fire as a consequence of which the Jewish lawgiver left behind to his people: 'The Lord your God is a devouring fire'" (I/8, 230). Schelling then locates the all-consuming ground of revelation in many other ancient sources. It is Heraclitus' inexhaustible fire (ἀκάματον πῦρ), whose quenching makes possible the creation of the cosmos (I/8, 230; Diels, fragment 30). It is the fiery winds, with their creatures and the wheels within wheels that visited Ezekiel (Ezekiel I: 13–16). It is "the object of the ancient Magi teachings," that is, of the Zoroastrian priests who taught that fire was the only possible guise of the supreme being (I/8, 230). This fire, if we here supplement Schelling by recalling the Mahāyāna tradition, is the emptiness of form, indwelling in form as the contradiction of form. As Avalokiteśvara, the sovereign bodhisattva who looks down and beholds the sufferings of the world, counsels Śāriputra in the *Heart Sutra*: "form is emptiness, emptiness is form; form is nothing but emptiness, emptiness is nothing but form; that which is form is emptiness, and that which is emptiness is form."[3]

In light of these examples, we can already say that fire appears in contesting the very forms that its ground had enabled, as if ground, without losing its unity, were nonetheless divided against itself. Just as fire cannot appear in a vacuum but can only appear in its act of consuming being, it is critical from the onset to insist that the opposition of fire to form is not a mere logical contradiction, as if we could have one or the other, but not both at the same time. It is rather the unity of contradiction, in which the opposing forces (form and emptiness, being and fire, or what Schelling calls A^1 and A^2) are two dynamic, interdependent forces of the same underlying force (what Schelling calls the A^3). These opposites, Schelling had insisted early on, form a curious "identity," but not because we are absurdly speaking of an equivalence of opposites (difference = identity), as if they really were one and the same thing (*einerlei*). Their "identity" is the unity of the opposition of difference and identity. This unity is a "difference" beyond and within the duality of difference and identity. When palpating identity, one finds difference, and when palpating form, one finds the fire of emptiness, yet these two potencies belong together. At the same time, emptiness, even though it contests form, is nowhere to be found independent of form. Given that Schelling wants to speak of the consuming fire of God, its solitude beyond any possible appearance, its chaos contesting appearance as such, it is a perverse theological fantasy that the solitude of God is found somewhere else than in the solitude of all that appears, each solitude maintaining the singularity of its appearance.

We could again supplement the movement of Schelling's thinking by remarking that this is also the curious "identity" that the great Shih-t'ou (Jp. Sekitō, 700–790 ACE) proclaims in the *Sandōkai*, a poem that is still chanted in Zen practice. The title speaks to the harmonious belonging together of difference of identity, which plays out in the body of the poem: "In the light there is darkness, but don't take it as darkness;

in the dark there is light, but don't see it as light. Light and dark oppose one another like the front and back foot in walking" (ZS, 33).

It is not surprising that a half millennium later the great Dōgen Zenji would insist that "mountains walk." The unity of walking, or, we could say in Schelling's idiom, the progression of the potencies, does not subsume everything under something greater, thereby evacuating each being of its haecceity.[4] The solitude of each walking, the shared singularity of each being, leads Shih-t'ou to conclude, "each of the myriad things has its merit" (ZS, 33). For Schelling, this is to say each being is singular, each being is born of solitude, because each being partakes in the paradox of coming to presence: each and every coming to presence, what each being shares in its own way, is therefore a solitary coming to presence. Each being is exposed as singular, or, as we saw in the previous chapter with Schelling's adaptation of Leibniz's monad, it is the very figure of shared solitude because each monad "is a particular that is as such absolute" (I/2, 64).

Solitude is not individuality. It is not that I am mathematically one and secondarily two. Nature is not a community or class of individual natural entities. There can be no merely individual solitude because singularity is the partaking of the interdependence of nature. The "community," the co-appearance, that is nature is the strange, fiery, unprethinkable *one* expressing itself as the irreducibly singular proliferation of the *many*, much in the way that Jean-Luc Nancy claims that the "world has no other origin than this singular multiplicity of origins" (BSP, 9).

In *The Ages of the World*, Schelling calls this solitary fire within the walking of all things, this saturnalian gravity within the coming to presence of light itself, the Godhead (the God beyond the being of any actual God), "something inaccessible to everything else, an irresistible ferocity, a fire in which nothing can live" (I/8, 299). Fire cannot sustain its own being, but rather feeds on that which has being, consuming it, taking it away from itself. "Panthers or tigers do not pull the carriage of Dionysus in vain. For this wild frenzy of inspiration in which nature found itself when it was in view of the being was celebrated in the nature worship of prescient ancient peoples by the drunken festivals of Bacchic orgies" (I/8, 337).

What does it mean to speak of the Dionysian panthers and tigers as the ground of nature? *Such a ground consumes what it produces.* Indeed, it first announces itself not in its productions, not in the creativity of nature, for nature, as we saw in the previous chapter, initially comes to the fore as regular, necessary, formal, mechanical, as if Newtonian mechanics vouchsafed nature's analytic frigidity. The creativity of nature, the sovereignty of its ground, must first break through the primacy of nature's necessity. As Schelling already asked in the 1797 *Die Weltseele*, "How can nature in its blind lawfulness lay claim to the appearance of freedom, and alternately, in appearing to be free, how can it obey a blind lawfulness?" (I/6, ix). The natality of nature, that nature is free for new beginnings, that its transitions are not deducible or otherwise knowable in advance, that it is not a closed, thoroughly recursive, internally self-regulating system of laws, paradoxically first announces itself in fatality, in the auto-consumption of nature, in death, in the fire of the ground, in the chariot of Dionysus dissolving the mechanical tyranny of form. Death, like fire, feeds on life, dissipating its forms. Death is withdrawal, decomposition, disintegration, what-cannot-be consuming what-is, but as such, it reveals itself within the ground of being. As Georges Bataille once articulated it: "The sacred is exactly comparable to the flame that destroys the wood by consuming it."[5]

It is not only revealed through death, but also through sin, that is, through the sudden awakening that we have already constitutively betrayed the ground that has unexpectedly broken through the repressive mechanical lawfulness of the ordinary mind. Although we did not intend to have missed the mark, we nonetheless accept responsibility for our actions and for our manner of being, much like the tragic hero affirms their misbegotten fate at the end of Schelling's *Letters on Dogmatism and Criticism*. Nonetheless, what is revealed by way of its consumption of form cannot be said to have any being of its own, not even the being of ground. The ground gives way in death, as well as in guilt consciousness. The latter here suddenly appears as the flight from ground that is always already accomplished such that the fiery ground, like death, "appears" like a thief in the night. It no longer regulates being, but appears as the abyssal source of regularity, the miracle, if you will, of the lawful. Schelling calls this abyss the originary unruliness of matter, its primal chaos and he likens this all-consuming fire to Plato's χώρα. This suddenly revealed, heretofore hidden other beginning, is not itself a being, but rather the possibility of there being anything at all, the incessant big bang by which there could be something rather than nothing, the beginning that makes all beginnings possible, but which in itself never ceases to begin anew.

Its revelation in withdrawal, however, also drives Schelling to concede that this is all, in a sense, a manner of speaking, a word that allows the silence of the ground to be heard in the form of the spoken.[6] Schelling's silent-word, so to speak, is the unity in the word of consonants and vowels, silence and sound. As he says this word in *The Ages of the World*: "The forces of that consuming fire still slumber in life, only pacified and, so to speak, exorcised by that *word* by which the one became the all" (I/8, 268). Schelling's enunciation of the silent-word resonates with the Mahāyāna tradition's remarkable deployment of language.

In the *Vimalakīrti Sutra*, for instance, we learn that wisdom is the mother and *upāya* is the father. *Upāya*, skillful means, is the capacity to communicate wisdom in accordance with the idiom and mind-set of one's interlocutors. It is the capacity to translate the Great Mother of wisdom; and wisdom, without *upāya*, is mere bondage. Detached from language it cannot say itself. The word for wisdom, *prajñā*, in the Mahāyāna tradition, contrasts pointedly with knowledge, *jñāna*. It is not a body of knowledge of any kind, nor is it communicable through discursivity. The prefix *pra* indicates anteriority, that which precedes *jñāna*. In Greek one would say *pro-gnōsis* (γνῶσις), but with the strong caveat that prognosis is typically the art of predicting the future, of speaking for the future, rather than allowing the future to speak for itself. One might say that *prajñā* is at the ground of all that might be knowable, but in itself it is unknowable. It is the preformal or empty possibility of form. *Upāya* translates *prajñā* into the word that can express what cannot say itself. The Buddha called this word, the word that skillfully carries the great sea of *prajñā* within it, the lion's roar, the word beyond all humanity that brings us into the friendship that is our being as nature.

On the one hand, the silence at the heart of that word allows the word to hold together the silence of the beginning and the clamor of the immense luxury of what has already begun. As such, it is a word like "nature" or "God," Schelling's own reinscription of Spinoza's *Deus sive natura*, and such words name this horizon of fatality and natality as a whole. On the other hand, this silence can be isolated in thinking (although not in God or nature) so that it emphasizes the primal chaos at the heart of this horizon. As such, this "word" is words like Godhead, silence, chaos, the abyss, the χώρα, the consuming fire, primordial nature (*Urnatur*, what in nature is oldest in nature), the tigers and panthers of Dionysus.

At the heart of the word, revealed in the collapse of the word as well as the miracle of the word, is the Godhead (the God beyond the being of God, the silence beyond the silence at the heart of the word). Having no being of its own, the Godhead can only be heard in the unity of the progression of potencies. It cannot be heard as *something to hear*. In this respect, following Böhme, Schelling also names the Godhead that appears in the decomposition of ground the *Ungrund*—for it is the moment in which decomposition has become revelatory, revealing itself as both ground and the decompositional contestation of all being, even the being of ground. The chaotic ground of ground, so to speak, is the infinite lack of being anything at all. It infinitely exceeds all specificity. The *Ungrund*, primordial nature, what is already too late to have happened, does not itself admit of composition. "It must be before every ground and before everything that exists and therefore must be a being [*Wesen*] *before* any kind of duality whatsoever. How could we call it anything else other than the primordial ground or better so the *nonground*?" (I/7, 406).[7] It haunts even the vertiginously abyssal experience of ground not just as the infinite depth of death and chaos, not just as the nonphilosophical at the heart of the philosophical, but also as the infinite fecundity of the barbarian principle, the "irreducible remainder" that can never be resolved by reason (I/7, 360), the could-be-anything-but in-itself-counts-as nothing, the pure *Seinkönnen*, at the heart of the ground of what-must-be.

This is emphatically not the claim that there is something grand, some great sacred Other, out there beyond all things or at the bottom of all things. As Deleuze and Guattari have rightly admonished: "Whenever there is transcendence, vertical Being, imperial State in the sky or on earth, there is religion; and there is Philosophy whenever there is immanence."[8] Schelling, after all, is not surrendering the autonomy of philosophy to the institutions and doctrines of any established church. He is arguing for *philosophical religion*. It may be that, as Deleuze and Guattari argue, "the first philosophers still look like priests" (WP, 43), and it is difficult to deny that Schelling, perhaps like Spinoza, but certainly unlike Nietzsche, consciously opts for a lexicon that had been historically co-opted into the authoritarian defense of a grand transcendent object, to which all of creation and all thinking, is subservient.

Such subservience is the impossibility of solitude. In this sense, it might be apt to declare Schelling (and Spinoza before him) a thief. Philosophy, long subservient to the autocracy of religion, takes these words back as philosophical words. As such, it is no longer the case that we are first religious and we then ask philosophically what this entails. In this noble heist, it is the case that we are first philosophical, and thinking discovers that it has a religious dimension. Such a dimension, however, asks that we think quite philosophically about what it would even mean to be religious or to speak of the future of God. That is, it would hear the thought of the future in the word "God."

Perhaps it could nonetheless also still be said that such language continues to flirt with the humiliation of solitude and the degradation of philosophy to the maintenance of the "immanent *to* Something," but I think that is not to read Schelling attentively just as it is not fair to do the same to Spinoza, a thinker whom Schelling deeply admired and whose vital impulses he sought to unleash. Spinoza's *Deus sive natura* is not a reactionary reinstallation of the very vertical Being that tormented him. As Deleuze and Guattari insist, "What singles out the philosopher is the part played by immanence or fire. Immanence is immanent only to its itself and consequently captures everything, absorbs All-One, and leaves nothing remaining to which it could be immanent" (WP, 45).

In this sense, Gianni Vattimo is right to sense that Schelling, as Schelling himself acknowledges in the Philosophy of Mythology and Revelation lectures, is a spiritual heir to the legacy of the medieval Calabrian prophet Joachim of Fiore who had intimated the incarnational coming of a third age, an age of the spirit, that is, an age of freedom. As Joachim of Fiore prophesied, following the apostle, "Where the spirit of the Lord is, there is freedom."[9] This intimated age, Vattimo observed, is not an age in which some particular event will transpire, as if prophecy were confused with fortune telling. (Prophecy is the revelation of a future that really has a future.) The age of spirit "stresses not the letter but the spirit of revelation" (AC, 31). This spirit is the destruction of the sectarian spirit. It no longer divides the earth up into those who are with us and those who are against us. It is the age of friendship and it is born of the fundamental mood of charity and compassion. Perhaps one might say it is the age in which the revelatory character of revelation is revealed. In such an age, an age that seems compatibly presaged, Vattimo argues, by the defining character of our age, "the end of metaphysics" (AC, 31), the future is no longer foreclosed by our metaphysical desperations. As outrageous as it might sound, we could say that it is to prophesy Foucault's "new age of curiosity":

> Curiosity is a vice that has been stigmatized in turn by Christianity, by philosophy, and even by a certain conception of science. Curiosity is seen as futility. However, I like the word; it suggests something quite different to me. It evokes "care"; it evokes the care one takes of what exists and what might exist; a sharpened sense of reality, but one that is never immobilized before it; a readiness to find what surrounds us strange and odd; a certain determination to throw off familiar ways of thought and to look at the same things in a different way; a passion for seizing what is happening now and what is disappearing; a lack of respect for the traditional hierarchies of what is important and fundamental. I dream of a new age of curiosity.[10]

Foucault dreamed of the future, Joachim of Fiore, in his own Medieval context with its own hermeneutic situation, prophesied a future of the spirit in which the revelatory character of things redeems things in the singularity of their solitude, and therefore in the exaltation of their strangeness. Schelling, too, dreamed Joachim's dream:

> An author of the 17th Century, the famous Angelus, called Silesius, counts among his epigrams:
>
> Der Vater *war* zuvor, der Sohn *ist* noch zur Zeit,
> Der Geist *wird* endlich sein am Tag der Herrlichkeit.
> The father *was* before, the son still *is* at present,
> The spirit *will* finally be on the day of glory. (II/4, 72)

God is the potency of the past, and this is the great wisdom of the Old Testament. The son is the potency of the present and this is the great wisdom of the New Testament. The spirit, however, is not only to come, but the coming of the future as such. "A third economy, the advent of a third time, the time of the spirit . . . as Joachim figured" (II/4, 72).

To dream or prophesy the future is, in a sense, *to dream or prophesy that the future will one day have a real future*. Is that not, at least in part, what Schelling meant by his notoriously cryptic announcement at the beginning of all three early extant drafts of *The Ages of the World*? "The past is known, the present is discerned, the future is intimated. The known is narrated, the discerned is presented, the intimated is prophesied" (I/8, 199).[11] The intimation of the future speaks of *ahnen*, to have an inkling, a foreboding, a premonition. As such, to have an inkling of the future as such is not to have a premonition of *something*, of a particular thing. That of which we have had an inkling or intimation is itself *geweissagt*. The antiquated verb *weissagen* is prevalent in Luther's translation of the Bible, handling a verb that would now be more familiarly translated as *prophezeien*, to prophesy, from the Greek προφήτης, to speak for the gods, whether such speaking be oracular, as in the Greek cults, or a translation of the Hebrew *navi*, the mouthpiece of God.

Prophecies were traditionally the realm of dreams, visions, and ecstasies, but when Schelling exclaims that "Not only poets, but also philosophers, have their ecstasies" (I/8, 203), he did not mean that philosophers sought some kind of narcotic escape route from nature in the dark night when everything was everything else (the fantasy of freestanding chaos, the confusion of the solitude of God with the independence and transcendence of the grand object God). He certainly did not mean that one could say in advance of the future what the future will have been. The fires of the future reveal the futurity of the future, not future events. In the first draft (1811) of *Die Weltalter*, Schelling succinctly defines the prophet as the one who can discern the manner in which the past, present, and future hold together as a dynamic whole, the one who "sees through the hanging together of the times [*der den Zusammenhang der Zeiten durchschaut*]" (WA, 83).

Schelling insisted that there is no short cut. "We do not live in vision. Our knowledge is piecemeal, that is, it must be generated piece by piece, according to sections and grades, all of which cannot happen without reflection" (I/8, 203). To become the mouthpiece for the future, to utter the word that reveals the future, demands that one not lose one's way in the consuming fire. If one "speaks immediately from vision, then they lose their necessary standard and are one with the object. . . . For this reason, they are not a master of their thoughts and struggle in vain to express the inexpressible without any certainty" (I/8, 204). To prophesy is not to speak from vision, for that would also mean that prophecy also subordinates philosophy to religious vision. It is the piecemeal path in reflection through what in itself is a unified progression in order to reveal in language the divinity of futurity. In a sense, we can say the prophet brings the darkness to light, reveals the darkness in the light and the silence in the sound. As such, we might even say the voice of the prophet, enunciating the premonitory fires of fatality and natality, calls for the coming of the kingdom of God on earth, for friendship with the solitude of being's inexhaustible singularities, for the day when the lamb lies down with the lion, for the coming of the Buddha lands. It was not the fault of the moon that the blind person did not see it.

We prophesy our intimations, our dreams, if you will, that the revelation of revelation will be the curiosity that the friends of wisdom in an age of spirit will take in all things in the nonmundane dignity of their strangeness. In this sense, in the intimation of an age of spirit, we can agree with Nancy when he writes that "'Nature' is also 'strange [*bizarre*],' and we exist there; we exist *in* it in the mode of a constantly renewed singularity" (BSP, 9). We speak to the intimation, to our *Ahnung*, that the future will one day have a future. This is explicitly not, returning to Vattimo, to reject an incarnational sense of temporality, to banish God to an absolute future, to think of God's absence exclusively, that is, dualistically, as an absence. The future is absolutely other than the present just as fire is absolutely other than form, but together they are the unity of progression, as we have said, a unity that is not entitative. God has no being, but cannot be an issue for being without partaking in being. A God banished to the vacuum of utter alterity reveals nothing, has nothing to say, gives rise to no words of silence. Vattimo: "God as the wholly other of which much of contemporary religious philosophy speaks is not the incarnate Christian God. It is the same old God of metaphysics, conceived of as the ultimate inaccessible ground of religion (to the point of appearing absurd) and warranted by his eminent objectivity, stability, and definitiveness—all traits belonging to the Platonic *ontos on*" (AC, 38). In this sense, the death of God, including the death of the God banished into the impersonality of absolute alterity, demands immanence and is vigilant that the great transcendent object does not once again sneak in the back door. The inkling that the future might one day have a future is the call to friendship in the strangeness of the earth's redeemed solitude.

In this sense, we can begin to appreciate why Schelling calls nature a "system of freedom," and why such a word does not absorb God into the order of the mundane. *Au contraire*, it reconfigures the mundane on an infinite plane, a plane in which Being itself emerges against the decompositional and recompositional background of infinity. As such, it is not the case that we first have religion than we have philosophy. If that is the case, we have only religion and philosophy as its police department. It is rather philosophy that one day awakens to the unexpected revelation of the heretofore-repressed infinity of nature. This is very different than the "religious authority" that "wants immanence to be tolerated only locally or at an intermediary level but on the condition that it comes from a higher source and falls lower down" (WP, 45). The barbarian Godhead emerges in this respect as "a backwards process . . . that, as always, ended with the consumption (by fire) of the previous idols [*Gebildete*]. . . . But that life, because in itself it is immortal and because it can in no way be, always again revives itself anew out of the ashes, like a Phoenix, and hence the eternal circle emerges" (I/8, 251).

The fires of nature's infinite plane of immanence, however, produce a curious paradox. The vertical structure of traditional religious life, in which the human surrenders her free will to a higher source, is not, as we have long imagined, to act naturally (in accordance with one's true being, to act in accordance with our natures, to be what we were meant to be). In a startling sense, the fantasy of natural law is sinfully irreligious, a betrayal of the Godhead that first emerges in the consuming fires of finitude, just as Virgil, in Hermann Broch's revelatory novel, *The Death of Virgil* (1945), realizing that he is on the verge of his own death, awakens to the revelation of his own betrayal of the elemental humus of being, the "the unborn state of the pre-creation." He who secured his glory by securing in his great epic the glory of the aspiring Roman *imperium*, suddenly realizes that all of this had been an immense perjury: "It was to await him, the unknown god, that his own glance had been compelled earthward, peering to see the advent of him whose redeeming word, born from and giving birth to duty, should restore language to a communication among men who supported the pledge." And this pledge, broken again and again, was a two-fold duty won in the face of the silent depths of the earth: the "duty of helpfulness, the duty of awakening."[12]

Virgil, who had dedicated his life to the poetic word, suddenly undergoes, via the impending decomposition of his own death, his coming elemental disintegration, an awakening, if you will, from the somnambulant sinfulness of his artistry and pseudoreligiosity. I want to use these words cautiously. Virgil died before the birth of Jesus of Nazareth and sin was certainly not part of his lexicon. Nonetheless, he realized that the *Aeneid* and its narrative of implacable destinies knew nothing of the deathly darkness of the Underworld. The dead do not command new empires and counsel new programs. The dead are not bureaucrats nor are they imperialists.

Virgil's poetic dead, the dead that Aeneas visits, hearkening back to Odysseus' visit to the Underworld, were simply the ruse of metaphor and his world vouchsafed the vertical Being of gods and emperors, words that surrendered the sovereign ground revealed by death to the orderly light of the understanding; and "life was only to be grasped in metaphor, and metaphor could express itself only in metaphor; the chain of metaphor was endless and death alone was without metaphor" (DV, 357). Virgil's masterful empire of metaphors perjured both the unknown God, and the preconscious pledge to friendship in an age of the awakening ἀνάμνησις of curiosity.

Broch often uses the leitmotif of suddenly waking, of regaining consciousness, from the initial somnambulance of human life. Schelling's attractiveness to psychoanalytically oriented thinkers like Žižek[13] suggests the sudden emergence of what is otherwise repressed. In the *Freedom* essay, the repressed barbarian beginning can suddenly and utterly unexpectedly spring forth. Perhaps one is "seized by dizziness on a high and precipitous summit" and "a secret voice seems to cry out that one jump." Or perhaps it is like that "old fable" (*The Odyssey*) in which "the irresistible song of the Sirens ring out from the depths in order to attract the passing sailors down into the whirlpool" (I/7, 382). As Lore Hühn has argued, the sudden flash of what is concealed in the self-enclosed self indicates "the structural model of the return of the repressed."[14] And what is repressed, even in the properly religious person? What does the security of a properly calculated and tidy religious life hide? Could it be the same deep-seated delusion that, as Zarathustra observed, allowed even the criminal the contentment of sound sleep? "Have you ever seen imprisoned criminals sleeping? They sleep peacefully, enjoying their new security."[15]

The flight from oneself (human freedom) to oneself (evil, sickness) is the perjury of the flight from the ground in the simultaneity of its fatality and natality. Although *Deus sive natura* cannot reverse existence and ground, they are separable in humans and the anguish of the *Urgrund/Ungrund* makes possible the flight to the periphery of existence. This is the freedom that in the *Freedom* essay drives the human in *Angst* to shun freedom—a freedom at the ground of the human that appears to each particular human will as "a consuming fire"

(I/7, 382), as a destructive contradiction to what one recognizes as the particularity of one's own life, with its habits, ambitions, and preferences. Humans abandon their freedom when their existence (which was in nature peripheral) becomes the ground of existence itself. This flight exploits the separabilty of ground and existence in the human by reversing them. To be clear: in sin, which we did not consciously commit, *the existence of consciousness takes its existence as its very ground.* The cogito innocently, that is to say, oblivious to the reversal that made it possible, surveys itself and its world. The cogito's self-awareness as a cogito was not a conscious deed, although its consequence defines the very fabric of consciousness as an experience of life and nature as grounded in oneself. When Schelling charges all of modern philosophy with having committed nature-cide because "nature is not present to it" and because "it lacks a living ground" (I/7, 361), this is nothing less than to say that modern consciousness is guiltlessly self-absorbed, that it arrives, even in natural science, as already fallen from nature. It is, so to speak, the unnatural quality of the so-called natural attitude.

The spell of ordinary, fallen consciousness (the natural attitude) can only be shattered in the becoming conscious of guilt for having sinned against freedom, a guilt that has nothing to do with assuming responsibility for one's conscious choices, for the reversal of ground and existence that is sin is precisely a reversal that enacts the seeming innocence of the fabric of ordinary consciousness life. It is not at all conscious of what made its very mode of consciousness possible. It is innocent even to the extent that the fact of its innocence would sound like an absurd accusation. It would find this essay utterly eccentric because it cannot access its own originary eccentricity. If consciousness itself is topsy-turvy (existence is the ground of ground), what is right side up appears upside down and what is upside down appears right side up. Only the gift of guilt allows consciousness to confront the perjury of consciousness and restore existence to the periphery of the humus of nature. In *The Ages of the World* Schelling claimed, "All life must pass through the fire of contradiction" for the latter is what is "innermost in life" (I/8, 321). The fire that consumes the bush, the fire that consumes the particular will, the fire at the heart of nature, is somehow also the life of nature and the treasure of human freedom.

In the Preface to the 1804 *Philosophy and Religion* essay, which inaugurates this development in Schelling's path of thought, he insisted that the divine fire of life is quite remote to the religious and philosophical world of the typical German, who, on the contrary, is given to the corporate fanaticism of *Schwärmerei*:

> For the Germans swarm around anything (*über alles gerathen die Deutschen in Schwärmen*), just like drones, but only to the extent that they try to carry away and consolidate something that was produced and that bloomed independently of them. But if they do make the effort to have their own thoughts, for which they are then themselves responsible, and if they stop pushing the responsibility for their own thoughts onto others, it would of itself hold back from them the cheap remark that they would completely explode if they had any thoughts of their own, for they are already so full to overflowing with foreign property. Furthermore, we leave them the periphery, but as to what concerns the internal:
> *Don't touch, old goat!—It burns*. (I/6, 15)

One might say here that Schelling is warning against surrendering the fires of one's solitude. Recalling Luther's original use of *Schwärmgeister*, of swarming, sexless bees, slavishly serving a queen, such smarmy, swarming, besotted fixations, are, already anticipating the *Freedom* essay, on the periphery of life. And what of this center that I shun by simply deriving my thoughts and values from the thoughts of others, from foreign properties? Could this ever have been the perjuring call from solitude to the nostalgic comforts of the return home? "*Don't touch, old goat!—It burns.*" We who would drone on in the irresponsibility of the periphery are the symptoms of *die Angst des Lebens* (I/7, 383).

The Future of God

In a sense, Schelling's audacious experiment is the endeavor to think the future of God, that is to say, the possibility of God having a future. Such a future does not reduce to speculating whether or not in the future we will be entitled to speak of God, to make reference to God. What can we say tomorrow of God? What can we, sobered by the death of God, by the putrefaction of the theological comforts that once spoke directly to our aspirations for ourselves and for nature itself, still say in today's spiritual climate? I am not a sociologist nor am I concerned with prognostication. I am concerned, however, with thinking the future of God as the demand that we not to exclude even God from the problem of time as such. In *The Ages of the World*, Schelling speaks of the temporality of revelation:

> These are the forces of that inner life that incessantly gives birth to itself and again consumes itself that the person must intimate, not without terror, as what is concealed in everything, even though it is now covered up and from the outside has adopted peaceful qualities. Through that constant retreat to the beginning and the eternal recommencement, it makes itself into substance in the real sense of the word (*id quod substat*), into the always abiding. It is the constant inner mechanism and clockwork, time, eternally commencing, eternally becoming, always devouring itself and always again giving birth to itself. (I/8, 230)

Time in its auto-consumptive potency was, for example, revealed to Arjuna when he beheld Krishna as *Kālá*, time, the world destroyer (XI: 32). In India, *Kālá* survives to this day in its female revelation as Kālī, the great destructive mother, prompting Sri Aurobindo to insist that we dare confront "not only of the beneficent Durga, but of the terrible Kālī in her blood-stained dance of destruction and to say, 'This too is the Mother; this also know to be God; this too, if thou hast the strength, adore.'"[16] Schelling, in *The Ages of the World*, reflects that "If we take into consideration the many terrible things in nature and the spiritual world and the great many other things that a benevolent hand seems to cover up from us, then we could not doubt that the Godhead sits enthroned over a world of terrors" (I/8, 268).

The world of terrors, however, is only part of the wheel of time, the wheels within wheels that Ezekiel's dream intimated. Their movement, in the perjury of the human's flight from freedom to the periphery of the self, is the awakening of the responsibility, the revelation of the pledge that stands outside of the seemingly mechanical spell of time. As this spell is broken, the Godhead reveals itself beyond Being yet within Being not only as what is oldest in nature, but as the future of nature. It is again important to insist here that this is not a reactionary insistence on the verticality of God. As Jean-Luc Nancy has argued, the "distinctive characteristic of Western monotheism is not the positing of a single god, but rather the effacing of the divine as such in the transcendence of the world" (BSP, 15). For Nancy, the very notion of creation, which for Schelling is critical to the very idea of nature, "contributes to rendering the concept of the 'author' of the world untenable. In fact, one could show how the motif of creation is one of those that leads directly to the death of God understood as author, first cause, and supreme being" (BSP, 15).

The death of God is the awakening from the repressive slumber of the verticality of being. In evil, for example, it is I, fleeing the world of terrors, that either imagines that I am God or that I am simply myself. In the first scenario, played out powerfully, for example, when Augustine realizes that he wanted to be God, the self in evil subsumes everything to itself. The evil will, left unchallenged, consume everything that is not itself and, in the end, it affirms solidarity with no thing and no one but itself; in the end, there will have been no world that is not my world. Evil in its imperiousness recognizes nothing but itself without end and would do so until there is nothing left to recognize. In the second scenario, in melancholy and depression, I bemoan the fact that I am only myself. As an immense and abyssal lack, I am never enough for myself, paradoxically hating myself for being myself. In awakening I turn to the ground right in from of me, otherwise than me, yet still caught up in me. As Nancy articulates it, "'God' is itself the singular appearance of the image or trace, or the disposition of its exposition: place as divine place, the divine as strictly *local*" (BSP, 17).

But animals and rocks do not know God. In a sense, "God" does not belong to any kind of knowledge, nor does God magically appear to thinking in faith if we imagine that the latter is a surrogate for thinking. Only humans, yes, alas, only humans, have God as an issue and in so doing are themselves at issue. This is true even for those to whom it will never occur even to think of something called "God," or to those for whom the traditions of speaking of and for God are repellent. "This thinking is in no way anthropocentric; it does not put humanity at the center of 'creation'; on the contrary, it transgresses [*traverse*] humanity in the excess of the appearing that appears on the scale of the totality of being, but which appears as that excess [*démesure*] which is impossible to totalize. It is being's infinite original singularity" (BSP, 17). When Schelling embraces guilt on behalf of all of nature, this is precisely to reject the centrality of human beings (this was the experience of evil as consciousness became the ground of nature). The claim is quite unexpected, given the dreary history of the elevation of the human over the nonhuman. In guilt, humans can embrace all of nature as the solitude, the being-singular-plural, if you will, of revelation. Nancy: "If existence is exposed as such by humans, what is exposed there also holds for the rest of beings" (BSP, 17).

This does not, however, bifurcate appearance into two hierarchically opposed camps: a privileged space of appearing for humans, and a dumb, starry-eyed prison of presence for everything else. Hence, in a line hard to appreciate and no doubt immediately off-putting to contemporary sensibilities, Schelling, borrowing from his Munich colleague and one time friend, Franz von Baader, insisted that humans can only stand below or above animals (I/7, 373). To stand above nature is not to flee nature nor is it to elevate the existence of humans as supreme. "On the contrary, in exposing itself as singularity, existence exposes the singularity of Being as such in all being" (BSP, 18).

Finally, it is important to say that revelation is not merely mythic. It is the overcoming of the merely mythic. This is not to distance myself altogether from recent accounts that recognize something mythic in Schelling's essay,[17] but, in the end, the *Freedom* essay is not the myth but rather the revelation of love: the conversion to the ground of existence as the affirmation of nature itself in all of its solitude. This is the revelation to humanity of the very humanity of humanity as the solitude of God as the shared solitude that extends throughout all of nature, that is to say, which is coextensive with Being itself. It is the simultaneous appearing together in the singularity of its solitude of all that is. "Existence, therefore, is not a property of *Dasein*; it is the original singularity of Being, which *Dasein* exposes for all being" (BSP, 18).

Love is born of the becoming guilty of consciousness in the dawning of its elemental perjury. If death is revelatory of the eternal beginning of life, life itself first demands the mortification (*Absterbung*) of the false life on the periphery. Life is only possible through what Zen called the Great Death, the dying to oneself in the coming to life as love. Hence Schelling warned that anyone "who wants to place themselves at the beginning point of a truly free philosophy, must abandon even God."[18] To be clear: *revelation does not begin first with the assumption of religion*. It is the moment in which the love of wisdom is the wisdom of love, its turning away from itself to the ground of nature. Such a return is only for one who "had once left everything and who were themselves left by everything" and who, like Socrates on the brink of his death in the *Phaedo*, "saw themselves alone with the infinite: a great step which Plato compared to death" (IPU, 18–19). Hence:

> What Dante had written on the gate of the Inferno could also in another sense be the entrance into Philosophy: "Abandon hope all you who enter here." The one who wants truly to philosophize must let go of all hope, all desire, all languor [*Sehnsucht*]. They must want nothing, know nothing, and feel themselves bare and poor. They must give up everything in order to gain everything. (IPU, 19)

If one wants to gain nature, one must first lose nature. If one wants to lose nature, one must lose oneself. Beyond all hope, one finds the solitude of God and the infinite singularity that is the being together of nature. This is philosophy's own "conversion" through guilt and mortification to the religiosity of nature's immanence, to *Deus sive natura.*

Philosophical Religion

In his still poorly understood turn to the philosophy of mythology and revelation, Schelling does not introduce a reactionary Christianity in order to recoil from the problem of revelation. Schelling had no interest in orthodoxy and did not associate Christianity with a doctrine (*Lehre*), but rather considered it a matter for thought (*Sache*) (II/4, 228). Schelling's lectures explicitly did not seek to "establish any actual doctrine or any kind of speculative dogmatism for we will not in any way be dogmatic" (II/4, 30). At issue is not any institution's promulgation of any particular doctrine of revelation. "The matter for thought [*Sache*] of revelation is older than any dogma and we will simply occupy ourselves with the matter for thought and not with any of the various ways it has been subjectively conceived" (II/4, 30).

This did not mean that Schelling went to war against established or "understood" religion—he attempted to maintain a healthy respect throughout. This is nonetheless a radical genealogical retrieval of the possibilities of religion. Schelling associated the historically domineering (universalizing by way of the sword) Catholic Church with Peter, and just as Peter wept tears of remorse for betraying Christ, one day the Catholic Church will do likewise. Although the Catholic Church possessed the mystery, they did not understand it and were given to "abiding, constant domination."[19] Indeed, "the pope is the true antichrist" (PO, 318) and the Petrine Church is "in the city" (U, 708), that is, is the Church of empire. If Peter is the official Church (A^1) and the hold of the *past*, then the Pauline Church, is the hidden Church (A^2) (II/4, 707), "external to the city [*in der Vostadt*]" (U, 708). Just as Dionysus in the mystery religions brought the real back to its soul (PO, 239), the Pauline retrieval of the esoteric dimension is the revelation of the *present* as grounded not in any particular thing or event, but abysmally rooted in the still creative depths. Not only does Paul challenge the Petrine Church, he foreshadows the power of the Protestant revolt. "In Paul lives the dialectical, limber, scientific, confrontational principle" (PO, 317). It is, however, only a "transitional form" (PO, 320), although "Paul's lightening strikes of genius . . . liberated the Church from a blind unity" (PO, 322) and revealed the ideal within the real. Protestants and their affirmation of knowledge counter the universality of the Catholic real, which was already contested within itself by its own mystical traditions. If Schelling were to,

> build a Church in our time, I would consecrate it to Saint John the Evangelist. But sooner or later there will be a Church in which the three above named apostles are united; this Church will then be the true be the true pantheon of Christian Church history! (U, 708)

The Church to come is the Church not merely *in* the future but *of* the future, that is, a church that liberates time and activates the creativity of the future, a church *of the whole of time*, a Church of the A^3. John is the "apostle of the future" (PO, 317): "One is not merely an apostle in the time in which one lives" and John was "actually not the apostle for his time." Paul would first have to come, but the "Lord loved" John, "that is, in him he knew himself" (U, 703). The Johanine Church, the church for everyone and everything—for all things human and for the mysterious creativity of the earth itself—is the "being everything in everything of God" (U, 708–709), a "theism that contains within itself the entire economy of God" (U, 709). This religion, what Bataille would later call "radical economy," not only excludes nothing, but also includes everything as alive, where "everything has its inner process for itself" (U, 710). This is religion beyond Petrine empire (the imposition of external forms of religion) and Pauline revolt (the recovery of the esoteric soul of religion). This, as we see in the final chapter, is the advent of a new mythology of Christianity and the liberation of a Christian creative ἀγάπη, the life of "a *revealed* God, *not an abstract idol*" (PO, 324). This is religion that artistically participates in the creativity of the universe's autogenesis.

Schelling concluded the first version of the *Philosophy of Revelation* lectures by remarking that he would have liked to have included some meditations on the "development of Christian poetry and art overall. It would have been remarkable—but the time does not permit it" (U, 710). Certainly, such art would no longer have been the dreary illustrations of the orthodox forms of religion. Indeed, this was the liberation of religion's sublimated creativity as well as the cessation of its bellicose domination of all opposing forms. But perhaps not only did the remaining lecture time not permit such considerations, but also the times *themselves*. As Schelling mused in the *The Ages of the World*: "Perhaps the one is still coming who will sing the greatest heroic poem, grasping in spirit something for which the seers of old were famous: what was, what is, what will be. But this time has not yet come. We must not misjudge our time" (I/8, 206).

Certainly, whatever its merits, pitching the problem of philosophical religion in exclusively Christian terms, despite Schelling's radical repurposing of them, remains problematic and to pursue this vision exclusively in such terms seems doomed to failure. Perhaps the project would now have to take up a language that emerges within a broader consideration of the immense variety of living religious experience all over the world. But then again, this is a question of philosophical strategy, of how best to speak in one's time of philosophical religion. Nonetheless, it is also worthwhile to appreciate how extraordinarily rare and generous Schelling's gesture of noncoercive and radically inclusive religiosity is:

> John is the apostle of a future and for the first time truly universal Church, this second, new Jerusalem that he saw descend from the heavens prepared as a bride beautifully dressed for her husband,[20] this no longer exclusive City of God (until that time there is an enduring opposition) in which Jew and Pagans are equally received. Without narrow coercion, without external authority, Paganism and Judaism are equally embraced for

> what they are and the Church exists through itself because each person voluntarily partakes in it, each through their own conviction that their spirit has found a homeland [*Heimat*] in it. (II/4, 328)

This is the kind of Jerusalem of which William Blake wrote and no doubt Schelling's words also have this kind of lonely imperceptibility. This "truly public—not state, not high church—religion" (II/4, 328) is the utopian promise nascent within philosophy, religion, and art. "It is the religion of humankind [*Menschengeschlecht*]" (II/4, 328), a "structure which encompasses everything human [*alles Menschliche umfassender Bau*]" and "in which nothing may be excluded and in which all human striving, wanting, thinking, and knowing are brought to consummate unity" (II/4, 296). This is no longer natural or mythological religion (A^1), the coming of the gods as themselves, nor is it a religion of revelation (A^2), but, finally, what Schelling called *philosophical religion* (A^3), a religion and a philosophy of and from the future: "this *philosophical religion does not exist*" and "it could only be the last product and the highest expression of the completed philosophy itself" (II/1, 250),[21] a widening and expansion, beyond mere rational considerations of supposedly religious objects, "of *philosophy and philosophical consciousness* itself" (II/1, 252; HCI, 175).

Part II

Thinking with Deleuze

3

Image of Thought

This Strangeness in My Life

It is hard to see where it is,
but it is there even in the morning
when the miracle of shapes
assemble and become familiar,
but not quite; and the echo
of a voice, now changed,
utterly dissociated, as though
all warmth and shared sweetness
had never been. It is this alien
space, not stark as the moon,
but lush and almost identical
to the space that was. But it is not.
It is another place and you are not
what you were but as though emerging
from air, you slowly show yourself
as someone, not ever remembered.

—Ruth Stone[1]

I begin this chapter with this beautiful and simple, but as such, deceptively easy, poem by the late American poet Ruth Stone. In the five meditations that follow, I let the poem act as a thread—perhaps that of Ariadne—to guide us through the labyrinth of the solitude of God. With this phrase, I bring together Schelling and Gilles Deleuze, especially his seminal work, *Difference and Repetition* (1968),[2] in order to posit the solitude of God as an image of Nature and to posit the latter as an image of thought, that is, as a prephilosophical horizon that allows *Naturphilosophie* to create, select, deploy, and arrange the concepts that belong to its manner of thinking. It is an image, however, like the vampire is an image: it cannot represent itself to itself speculatively, that is, catch itself in the speculum.[3]

It Is Hard To See Where It Is

What in the world could it mean to dare speak of the solitude of God with respect to Schelling? Such language generally embarrasses prevailing tastes and customs of thinking. Have we not at long last begun to shed ourselves of the shackles of such reactionary and hopelessly infantile philosophical commitments? Is this not therefore a feature of Schelling's thinking that we perhaps pass over in silence or treat as an antiquated relic in the mausoleum of the history of ideas?[4] This issue also comes immediately to the fore in my desire to use Deleuze's thinking to help reopen the space of Schelling's thinking. In his remarkable book on Nietzsche (1962), Deleuze quipped that "the Gods are dead but they have died from laughing, on hearing one God claim to be the only one."[5]

Thirty-two years later, he declared, in a manner befitting his allegiance to the image of thought exemplified by Spinoza and Nietzsche, that the "illusion of transcendence," perhaps the premiere mirage to which thinking is subject, renders immanence immanent *to something* and rediscovers transcendence lurking within immanence (WP, 49). God talk, subscription to a big something beyond thinking and nature, is the death sentence for philosophy.

Schelling, however, is no onto-theologian, nor does he sneakily pass the buck on the problem by taking cover in a "theology of mystery." Faith, Deleuze contended, "invites us to rediscover *once and for all* [*une fois pour toutes*] God and the self in a common resurrection" (DR, 95/127). Faith finds a way around the problem of the death of God and its corollary, the death of the anthropocentric subject. Deleuze approvingly turned to Blanchot, who argued that death does not reduce to a personal death. It is also, and more unsettlingly, the impersonality of death, death as what Deleuze calls "the insubordinate multiple" and what Blanchot called the "inevitable but inaccessible death," the "abyss of the present, time without a present" (DR, 112/148–149). Expanding on Blanchot, Deleuze argued that the "other death . . . refers to the state of free differences when they assume a shape which excludes my own coherence no less than that of any identity whatsoever. There is always a 'one dies' more profound than 'I die,' and it is not only the gods who die endlessly and in a variety of ways" (DR 113/149). Death, marking the absolute futurity of the future, tears asunder the realm of representation. Perhaps one could say that for Deleuze death is like Rilke's animals, who know that we are "not at home in the interpreted world." The interruption of death, which reveals the rift that representation occludes, shows us that once and for all that there is nothing that is once and for all.

So what does it mean to speak of the solitude of God, without recourse to either a (supreme) being *to* which all of immanence is immanent, or the possibility of a once and for all that is once and for all? Nietzsche claimed in book three of *Also sprach Zarathustra* ("Von den Abtrünnigen"), to which Deleuze's quip alludes, that the gods did not die in Wagner's gloomy twilight, but rather they laughed themselves to death when a jealous and old "furious beard of a god" declared (as in Exodus 20:3–5), "There is only one God. You shall have no other God before me." "At this all of the gods shook in their chairs from laughing and cried, 'Is this not precisely godliness that there are gods, but no God [*Ist das eben Göttlichkeit, daß es Götter, aber keinen Gott gibt*]?'"[6]

Nietzsche returned to this thought a couple of years later in *The Anti-Christ*, exclaiming "Almost two thousand years and not a single new god!" All of the other gods may have laughed to death, but the *one* God, the one that is single, individual, mathematically one, is not laughing. Its deadly gravity speaks to an image of thought that preserves the rule of the same by repressing and excluding (*determinatio negatio est*), inspiring Nietzsche to call out the dullness of the One that only is ever itself as the "pitiable god of Christian monotono-theism [*erbarmungswürdiger Gott des christlichen Monotono-Theismus*]" (§19; KSA 3, 183).[7] The God that (hopefully) died was the supreme being, eternally and seriously itself, incapable of surprise or of disarming its own identity in the conflagration of laughter. Godliness, on the other hand, was dynamically multiple, producing new gods, but exhausting itself in no god. Its a priori plurality resisted the monotony of representation, including the ever-repeating forms that stabilize religious life, giving it a single, deadly tone. Indeed, Deleuze speaks of his own empiricism as "a mysticism and mathematicism of concepts" (DR, xx/3).[8]

Oscar Wilde once retorted (in *De Profundis*) that Jesus was not about taking an interesting thief and turning him into a tedious honest man. In the same spirit, Schelling in the *Freedom* essay proclaimed, "God is not a God of the dead but rather of the living" (I/7, 346). But how is the solitude of God on the side of the living and not the side of monotony? (Death qua death is always only death.) This was in part Schelling's concern about Hegel: Hegel's God "is the God who only ever does what He always has done, and who therefore cannot create anything new; His life is a cycle of forms in which He perpetually externalizes Himself, in order to return to Himself again, and always returns to Himself, only in order externalize Himself anew."[9] Hegel's God spins in an ingenious circle, although it is the circulation of the identical as negativity.[10] Thinking finds itself oriented to its own form and the form of thinking repeats itself through the diversity of its content. In this sense, representation claims that it has attained the infinite domain of thinking while retaining representation. Deleuze called this "orgiastic representation," the final cunning of representation.[11] When thinking becomes a clear errand, that is, when it figures out its proper work, the seriousness of the concept, it forgoes genuine progress because it forgoes the insuperable rift that is the infinite depth of its past and the infinite distance of its futurity.

Such philosophies read as if written and populated by consummate beings, beings wholly obedient to themselves, reconciled to themselves, in which every tomorrow is simply another yesterday. Schelling called these obediently self-identical beings "angels," and he judged them in his *Kunstphilosophie* (1802–1803) as "the most boring beings of all" (I/1, 473). Angelic philosophy is the appropriately monotonous pursuit of Monotono-Theism.

Hence, it would not here make any sense to think of the solitude of God as an evocation of God as *sui generis*, a unique being whose species heads its own genus. Solitude does not mark the monotony of God. In the first draft of *Die Weltalter*, Schelling speaks of God's stultifying solitude: "For human beings are helped by other human beings, helped even by God; however, there is nothing whatsoever that can come to the aid of the primordial being in its stultifying solitude [*in seiner schrecklichen Einsamkeit*]; it must fight its way through chaos for itself, utterly alone."[12] God can be thought of in His *schreckliche Einsamkeit*, but the solitude of God's relationship to chaos does not imply that God is an individual. God is not a particular being, unique or otherwise.

God is singular, but in its singularity, it is the singularity of all things, that is, to borrow Jean-Luc Nancy's felicitous phrase, the being-singular plural of nature. To say that the solitude of God is the singularity of God, is not, however, to think singularity as any kind of One, but as the multiple, the plural. God's solitude, in the language of the third draft of *Die Weltalter*, is "the forces of that inner life that incessantly gives birth to itself and again consumes itself that the person must intimate, not without terror, as what is concealed in everything, even though it is now covered up and from the outside has adopted peaceful qualities" (I/8, 230).[13] There is nothing monotonous about God's solitude. God's terrible solitude is the "awful and the terrible" that lurks in the depths of Nature:

> If we take into consideration the many terrible things in Nature and the spiritual world and the great many other things that a benevolent hand seems to cover up from us, then we could not doubt that the Godhead sits enthroned over a world of terrors. And God, in accordance with what is concealed in and by God, could be called the awful and the terrible, not in a derivative fashion, but in their original sense. (I/8, 268)

The awful and terrible solitude of God is at the heart of *Naturphilosophie*, and my turn here to Deleuze is in the interest of finding new ways to articulate this problem. The task of this chapter is, in the end, to think *die Natur selbst* not as a concept, but as an image of thought. In turning to Deleuze,[14] especially his magnum opus, *Difference and Repetition*, which has some great affinities with Schelling, I am aware of a potential embarrassment. There is a painful irony in pursuing affinities between thinkers by turning to a book dedicated to overcoming the assumptions inherent in the resemblances that allow us to identify what is logically the same or similar. The ipso facto resemblance of Deleuze and Schelling is not only insufficient in itself to recommend thinking them together (what is the value of a mere coincidence?), but it also obscures the powerful critique of resemblance found in both thinkers. We leave Schelling and Deleuze to their respective singular multiplicities, but we hold them together in the more modest hope of appreciating anew the *Naturphilosophie* as an image of thought.

Speaking of the "branches" on which "difference is crucified," Deleuze argued that

> [they] form quadripartite fetters under which only that which is identical, similar, analogous, or opposed can be considered different: *difference becomes an object of representation always in relation to a conceived identity, a judged analogy, an imagined opposition or a perceived similitude.* Under these four coincident figures, difference acquires a sufficient in the form of a *principium comparationis.* For this reason, the world of representation is characterized by its inability to conceive of difference in itself; and by the same token, its inability to conceive of repetition in itself, since the latter is grasped only by means of recognition, distribution, reproduction and resemblance in so far as these alienate the prefix RE in simple generalities of representation. (DR, 138/180)

How can we otherwise hear the passage or movement of this prefix RE beyond its bare presentation again and again of the same? It is not to say that nature is whatever nature is, presenting itself again and again as the rule by which nature is always nature. Nature is not the presentation of the self-same nature in and through the history of its manifestations. Nature is not the ever re-presenting, the bare repetition or simple repeating of the same. For Deleuze, the intensity within nature that differentiates itself ever anew in all of its extensions "is the transcendental principle which maintains itself in itself, beyond the reach of the empirical principle. Moreover, while the laws of nature govern the surface of the world, the eternal return ceaselessly rumbles in this other dimension of the transcendental or the volcanic *spatium*" (DR, 241/310–311). Indeed, God "dances upon a volcano" (DR, 234/301).

Schelling for his part argued that "The idea of nature as exteriority implies immediately the idea of nature as a system of laws" (1/3, 6), and this was the legacy of modernity, but, as he argued in the *Freedom* essay, the representation of nature, even as the limit of representation as such, is the fatal flaw that characterizes modernity: "Nature is not present to it, and that it lacks a living ground [*die Natur für sich nicht vorhanden ist, und daß es ihr am lebendigen Grunde fehlt*]" (I/7, 361). Nature therefore becomes an abstraction; its forces become mere repetitions of the same. Natural laws are its inviolable operators, and nature is bereft of the miracle of natality, incapable of real progressivity, so that it merely repeats what it has always already been, "swiveling in the indifferent circle of sameness, which would not be progressive, but rather insensible and non-vital" (I/7, 345). Unless thinking illuminates the gap that allows one to think nature as the eternal beginning, nature occludes what is most forceful, most valuable, and most transformative within itself. It is repetition of nature's intensity, not the repetition of its laws, that shows itself in this gap, a gap that was at its heart. As Schelling posed the question in *Von der Weltseele*: "How can nature in its blind lawfulness lay claim to the appearance of freedom, and alternately, in appearing to be free, how can it obey a blind lawfulness" (I/6, ix)? For both thinkers, therefore, this is a question of thinking nature in its repetition, which is to say, in the productive temporality of its becoming, an issue to which we will later turn.

This is, however, explicitly not to present the question of nature as the *tertium comparationis*, the shared term or a quality of a comparison, the shared feature that both thinkers have in common. It could be said with some right that Schelling and Deleuze, each in their own way, share a commitment to the question of nature, but what they share in common demands paradoxically that we simultaneously move beyond relations of coincidence and resemblance. They powerfully resemble each other precisely in their shared commitment to displacing the ontological assumptions that enable resemblance. By having the problem of nature in common, they, each in their own way, partake in the solitude of nature and therefore also in the shared singularities of their own respective solitudes.

And the Echo of a Voice, Now Changed, Utterly Dissociated

Deleuze wrote no major treatise or essay solely dedicated to Schelling, and there is no evidence that Deleuze devoted considerable time to reading extensively Schelling's corpus. His familiarity with Schelling likely stems from the translations of Jankélévitch, namely *Essais* (Aubier 1946)[15] and a translation of the third (or 1815) draft of *Die Weltalter* (which included *Deities of Samothrace*) (Aubier 1949). Furthermore, Deleuze's preoccupation with Schelling more or less manifests in the period on which we are concentrating (1968), with allusions to Schelling in *Difference and Repetition* (1968), the first Spinoza book, *Spinoza et le problême de l'expression* (1968), and briefly in *The Logic of Sense* (1969).[16] With the exception of one issue, Deleuze clearly esteems Schelling's thinking, defending it from Hegel:

> The most important aspect of Schelling's philosophy is his consideration of powers. How unjust, in this respect, is Hegel's critical remark about the black cows! Of these two philosophers, it is Schelling who brings difference out of the night of the Identical, and with finer, more varied and more terrifying flashes than those of contradiction: with *progressivity*. Anger and love are powers of the Idea which develop on the basis of a μὴ ὄν—in other words, not from a negative or a non-being [*non-être*] (οὐκ ὄν) but from a problematic being or non-existent, a being implicit in those existences beyond the ground. The God of love and the God of anger are required in order to have an Idea. A, A^2, A^3 form the play of pure depontentialisation and potentiality, testifying to the presence in Schelling's philosophy of a differential calculus adequate to the dialectic. (DR, 190–191/246–247)

Deleuze, relying on the Jankélévitch translations of the *Freedom* essay and the 1815 *Weltalter*, clearly knew and appreciated Schelling's embrace of the decisive ancient distinction as it, for example, plays itself out in Plato's *Sophist*, between the negation or complete denial of being, οὐκ ὄν, and μὴ ὄν, that which in being is not of being. In the *Freedom* essay, Schelling takes up the question of the μὴ ὄν in relationship to what has been misconstrued as *creatio ex nihilo*. Creatures do not emerge out of thin air any more than they emerge from any material substance. Creatures emerge "εκ των μη όντων, out of that which *is* not there [*das da nicht* ist]" (I/7, 373). *Ex nihilo* misconstrues the "famous μὴ ὄν of the Ancients" (I/7, 373). Schelling made the point again in the 1815 *Weltalter*:

> They could have been liberated from this simple grammatical misunderstanding, which also prejudiced a good many interpreters of the Greek philosophers, and from which the concept of the *creatio ex nihilo*, among others, also seems to owe its origin, with this distinction, entirely easy to learn and which can be found, if nowhere else, certainly in Plutarch, between non-Being [*nicht Seyn*] (μὴ εἶναι) and the Being which has no being [*nicht seyend Seyn*] (μὴ On εἶναι). (I/8, 221).

Because the μὴ ὄν has been so often mistranslated as not-being (and hence nothing, lack of existence), it is associated with *nihil*—but if one can hear more carefully the μὴ ὂν repressed in our habitual associations of nothingness, then one could perhaps say that Schelling was a productive or creative nihilist in the sense in which Nietzsche latter spoke of "active nihilism."[17] The phrase μὴ ὂν names that which *is* as not having being, the *Ungrund* always paradoxically behind yet ahead of itself, the dark precursor that ruins the illusion of representation, the indivisible remainder contesting while supporting all beings. Schelling linked the μὴ ὂν with Penia, poverty, who, in Plato's *Symposium*, couples with Poros, means, to generate Eros, the generosity of their productivity. Contrary to the Neo-Platonic tradition, which reduces the miraculous poverty of the μὴ ὂν to an anemic sliver of a being that barely has any being, the μὴ ὂν, *der unendliche Mangel an Sein*, the eternal lack of Being,[18] is the inexhaustible *ewiger Anfang*, what in itself is always past yet still to come in what is present. As Schelling formulates it in the 1815 *Weltalter*:

> This extremity can itself be called only a shadow of the being, a minimum of reality, only to some extent still having being, but not really. This is the meaning of non-being according to the Neo-Platonists, who no longer understood Plato's real meaning of it. We, following the opposite direction, also recognize an extremity, below which there is nothing, but it is for us not something ultimate, but something primary, out of which all things begin, an eternal beginning [*ein ewiger Anfang*], not a mere feebleness or lack in the being, but active negation. (1/8, 245)

Deleuze, turning to Plato's subtle deployment of the μὴ ὂν in the *Sophist*, asks about the μὴ: the "'non' in the expression 'non-being' expresses *something other than the negative*" (DR, 63/88). That is to say, "being is difference itself" or better: "Being is also nonbeing, *but nonbeing is not the being of the negative*; rather it is the being of the problematic, the being of problem and question." Deleuze plays with three strategies to somehow convey the force of the μὴ ὂν. One could write it: "(non)-being" or better: "?-being." Or he links it to the French NE: "an expletive NE rather than a negative 'not.' This μὴ ὂν is so called because it precedes all affirmation, but is none the less completely positive" (DR, 267/343). It is the "differential element in which affirmation, as multiple affirmation, finds the principle of its genesis" (DR, 64/89).

Schelling for his part speaks variously, to cite some examples, of that within Being that does not have being, the eternal past as the negating force of the future, the *Überseyende,* God as the super-actual, beyond that which has being, and "therefore a sublimity beyond Being and Not-being," (I/8, 238) "the devouring ferocity of purity" (I/8, 236), and the second potency (the A^2). Schelling also refers to freedom by linking it to Plato's χώρα, that "wild, unruly matter or nature" (I/7, 326). These names, however, do not represent the silence that they nonetheless strive to express, and in a sense, Deleuze is right to insist that "repetition must be understood in the pronominal" (DR, 23/36).

That Schelling understands the μὴ ὄν, problem-being, as fundamental to *Naturphilosophie*, and that Deleuze should embrace this, should come to no surprise, given that both admire Spinoza. Although Schelling felt compelled to help Spinoza out with what he in the *Freedom* essay dubbed an interpretive "supplement" (I/7, 345), it was to free Spinoza from any of the dogmatic tendencies of modern philosophy and to unleash the powerful life that dwells in the gap that is the disequilibrium and vital disparity between *natura naturans* and *natura naturata*. Deleuze, in his first Spinoza book, sees this clearly, although Deleuze reads Spinoza as if no "interpretive supplement" were required: "Schelling is Spinozist when he develops a theory of the absolute, representing God by the symbol 'A^3' which comprises the Real and the Ideal as its powers."[19] Deleuze, however, despite Schelling's many defenses and developments of Spinoza,[20] does not attribute Schelling's "Spinozism" to Schelling's profound reading of Spinoza. Schelling derived his discourse on *Ausdruck* (expression) not from Spinoza, but from Jakob Böhme (EPS, 18). Although this claim is to some extent true, and the influence of Böhme on the *Freedom* essay is pronounced,[21] I suspect that it is an exaggeration to distance Schelling on this issue so far from Spinoza. (The *Freedom* essay also has a spirited meditation on Spinoza.) Moreover, I think that it is a mistake to attribute a rupture in Schelling's thinking that began with his discovery of Böhme. Ernst Benz's 1955 study, *Schellings theologische Geistesahnen*[22] demonstrated just how deep and thoroughgoing an influence the Schwäbian Pietists, the Rhineland mystics, and others were on Schelling's thinking from its inception. Bruce Matthews has recently shown that Schelling at an early age came under the influence of Philip Matthäus Hahn, writing a eulogy for him when Schelling was only fifteen. Hahn was dedicated to the "pursuit of the divine in nature, free of the ideological constraints of working in a university setting. The trajectory of his research precluded explaining Nature as a mechanism, demanding instead that nature be understood as the ongoing revelation of the divine life."[23]

Evoking Schelling's early experience of a radically incarnational *Naturphilosophie* is not, however, simply to set the record straight on Schelling's *Geistesahnen*, but it is also to begin to suggest the image of thought that led Schelling's concept selection and creation through Hahn, Plato (the "divine" one), a rethinking of Kant's entire critical project from the perspective of what opens up in the third *Critique*, Hamann (the *wunderlich* one, strange, mysterious, wondrous), Spinoza (with the interpretive supplement), Böhme, and others. "We head for the horizon, on the plane of immanence, and we return with bloodshot eyes" (WP, 41).

Almost Identical to the Space That Was. But It Is Not

Despite acknowledging their striking affinities, Alberto Toscano, in distancing Schelling from Deleuze and Guattari, spoke of Schelling in terms "of something like an *organic image of thought*."[24] It belongs to the dignity of difference that we leave each path to its own—it is enough that these affinities have the requisite potency to open in new ways the power of Schelling's *Naturphilosophie*. I do find, however, Toscano's phrase felicitous. *Naturphilosophie* is not a philosophy about nature. It is not a philosophical conception *of* nature. It is philosophy in the image of nature and reciprocally, nature as the image of philosophy.

As I develop in more detail in chapter 6, this is certainly not to suggest that by image we mean a depiction, illustration, or representation. Furthermore, Schelling's thinking does not articulate a particular position, not even that of "idealism." Schelling's thinking requires that the reader enter the *horizon* or *climate* of his thinking. One cannot simply extract arguments here and there and compare them with the decontextualized arguments of other thinkers. Schelling, to borrow the image of philosophy from Deleuze and Guattari, is conceptually articulating an infinite yet immanent *plane* of thinking. Working in tandem, relevant concepts strive to allow the prephilosophical ground of philosophy to emerge within philosophy. The prephilosophical plane of philosophical concept creation is not itself a concept or a phenomenon, but what they called a *planomenon* (WP, 35), "the horizon of events, the reservoir or reserve of purely conceptual events" (WP, 36). In other words, the fundamental image of thought dictates what by right belongs to thinking, what it should strive to articulate conceptually, and what it can ignore on grounds of relevance and taste. It inspires the philosopher's conceptual creativity and ingenuity.

I realize that this is perhaps a stretch, but I think that this implication is fully within the movement of Schelling's thinking: Nonthought is at the heart of *Naturphilosophie*. At the heart of philosophy is that which is always becoming philosophical, but which in itself is nonphilosophical, dissembling itself in its donation of all things philosophical. This is also exhibited in the vast range of disciplines that suddenly become philosophical—a vast expansion of territory and provocations thinking. The *Naturphilosophie* is a natural, or if you will, an *organic* image of thinking, a prephilosophical intuition of the horizon that shapes what belongs by right to this manner of thinking.

It seems to me that Schelling's organic image of thought, thinking as the creative unfolding of and within a living, ever recommencing, inexhaustible whole, is an image of the solitude of God, of stepping "onto the path of times" where God becomes "exclusively singular" (I/8, 312) as an image of thinking as such. Schelling even provides a differential calculus for this, that is, a calculus that does not repress difference, but rather releases the differential temporality and spatiality of the potencies:

> Hence the whole can be designated in the following illustration:
>
> $$\frac{A^3}{A^2 = (A = B)}$$
>
> This is the One and the Many (Ἓν καὶ πάν) in intimate connection. (I/8, 312)

It seems to me that the meeting of Schelling and Deleuze allows us to pose the question like this: Does an organic image of thought allow for the maximization of difference? Schelling's organic image of thought is not fettered to Hegel's strategy of orgiastic representation, of an infinitely large archive of representation.[25] The calculus of potencies unleashes an immense play of difference and therefore deeply multiplies philosophy's themes and problems. It is certainly a powerful antidote to the dogmatic image of thought in "good sense" and "common sense" in which "everybody knows and is presumed to know what it means to think" (DR, 131/172). Schelling has Nietzsche's courage for his own becoming against the tyranny of convention's alliance with uninterrupted presence.[26] *Naturphilosophie* allies itself with *terra incognita*.

Yet we must ask if the organic image of thought is up to its own test: Is there some manner of thinking that the organic image of thought excludes? Does it structurally repress under the question of nature what is nowise thinkable within the question of nature? Is nature *the* plane of immanence? Deleuze's thinking takes up Schelling, but along with many problems that Schelling may well not even have found recognizable. In the end, the sufficiency of an organic image of thought may be an insuperable problem. As Deleuze and Guattari admit with reference to their own plane of immanence: "We will say that THE plane of immanence is, at the same time, that which must be thought and that which cannot be thought. It is the nonthought within thought. It is the base of all planes, immanent to every thinkable plane" (WP, 59). This is not, however, to again lament the tribulations of finitude and persecute those who found the courage for their own becomings. "Perhaps this is the supreme act of philosophy: not so much to think THE plane of immanence as to show that it is there, unthought in every plane" (WP, 59). Perhaps it is sufficient simply to say that Schelling was dedicated to this supreme act of philosophy.

It Is Another Place

Schelling, as did Böhme, synthesizes the ground to its eternally antecedent *Ungrund*. "There must be a *Wesen* before any ground and before everything existent, and therefore altogether before any duality. How else can we name it than as the *Urgrund*, or better, the *Ungrund*" (I/7, 407)?[27] Deleuze synthesizes *le fond* (ground, *Grund*) in its appearance as "homogeneous extensity"—space that is simply space—with a "projection" of *le profond* (DR, 229/296), the antecedent to ground that is the depths and distances from ground. The illusion of homogeneous space is made possible by the heterogeneity of the depths.

Hence, Deleuze speaks of a "synthesis of depth," which "endows the object with its shadow, bears witness to the furthest past and to the coexistence of the past with the present" (DR, 230/296). The depth of pure space makes possible extended space, and in so doing, pure space is always and irretrievably *before* extended space, yet still to come in the temporal dynamism or passage of place, the infinite distance that is the eternal past and future of livid space. "Depth is the intensity of being [*la profondeur est l'intensité de l'être*], or vice-versa" (DR, 231/296). Depth is the rumbling in the earth that seemed to support our cities of representation. "Representation fails to capture the affirmed world of difference. Representation has only a single center, a unique and receding perspective, and in consequence a false depth [*une fausse profondeur*]. It mediates everything, but mobilizes and moves nothing" (DR, 55–56/78).

Schelling had long argued that gravity lay concealed in the depths of light, holding back its extensity, keeping it from settling into itself, yet just as light cannot flee the darkness of its ground, the ground cannot detach itself from what it contests.[28] In the *Freedom* essay, we find the remarkable formulation: "Gravity proceeds light as its eternally dark ground. Gravity is not itself *actu*, and it escapes into the night in which the light (the existing) dawns. Even light cannot fully loosen the seal under which gravity lies contained [*beschlossen*]" (I/7, 358). There is no independent gravity. It remains contained under the seal of light just as the ground of God lies in the depths under the seal of God's existence. The ground is stamped as nature, and nature cannot altogether loosen what it has formed. "Neither is the other and yet one is not itself without the other" (1/7, 359). Schelling calls this the "circle out of which everything becomes" (I/7, 359) and in the 1806 essay on gravity and light we find a remarkable expression of this circle:

> This same all-inclusive and appearing unity, which the movements of universal nature, the quiet and continuous as well as the powerful and sudden changes controls in accordance with the idea of the whole, and always returns everything into the eternal circle . . . is from the moment of its emergence determined to seek to change earth, air and water into living beings, images of its universal life. (I/2, 374)

This is a monstrous circle, the paradoxically revelatory withdraw of its gravitational center, a center that Schelling will also associate with the absolute spatial and temporal continuity of the χώρα. "The copula [*das Band*] that binds all things and produces the one in the many, the everywhere present, the placeless, un-circumscribed central point, exists in nature as gravity" (I/2, 364). This circle is not what we have marked as Hegel's circle of orgiastic representation, but rather the circularity of the eternal return of the same, that is, difference and its eternal repetition. In *The Ages of the World*, Schelling subtly but decisively argues this issue:

> This cision, this doubling of ourselves [*diese Scheidung, diese Verdoppelung unserer selbst*], this secret circulation in which there are two beings . . . this silent dialogue, this inner art of conversation, is the authentic mystery of the philosopher. From the outside this conversation is thereby called the dialectic and the dialectic is a copy of this conversation. When the dialectic has become only form, it is this conversation's empty semblance and shadow. (I/8, 201)

Although Hegel is not mentioned by name, Schelling's concern is clear. What is key is not the *form* of the conversation, not the winnowing of reflection by which formulas are produced (including the remarkable formula of the world in Schelling's own negative philosophy). Having won the form, it does not repeat itself ceaselessly and the meaning of each image that emerges is not, as well shall see in our concluding chapter, the formula by which emergence itself emerges. This would reduce all philosophy to negative philosophy (presentations of freedom at the heart of necessity and their consequent formulas). "The authentic mystery of the philosopher" cannot be viewed from the periphery ("from the outside"), but only from within the immense creativity of its life, from the positive production of its ceaseless haecceities, its unprethinkable singularities, which emerge, as art demonstrates, simply as themselves. The dialectic is born from the creativity of the center, and the *Phenomenology* is a work of stunning creativity and admirably rigorous reflection. Nonetheless, as a form, it is what we later call a "dead word," a *mere form of life as if it were the form of life as such*, but merely as a peripheral form, subject to the same ruin of all forms of life (see chapter 1), its staying force is the lingering force of ghosts (the returning of the dead), an "empty semblance and shadow." Hegel's dialectic profits on its ruin and is as such, despite all of its brilliance, the dream of a resurrection machine, a machine that refuses the ruin of death and that cannot be brought to the brink of absolute silence. The *Phenomenology* is, as has often been noted, thinking's *Bildungsroman* in which at the end of its adventure it comes to glimpse the form that, unbeknown to it, it has always had and will always have. As Jean-François Lyotard succinctly analyzes this kind of recursivity (the eternal return of the dialectic):

> But the beginning can appear as this final result only because the rule of the *Resultat* has been presupposed from the beginning. The first phrase was linked onto the following one and onto the others in conformity with this rule. But this rule is then merely presupposed and not engendered. If it is not applied from the beginning, there is no necessity of finding it at the end, and if it is not at the end, it wouldn't have been engendered, and it was therefore not the rule that was sought.[29]

Lyotard calls for "silences, instead of a *Resultat*" (D, 106), and Schelling, too, offers no final conclusions, no moment in which we have finally learned something once and for all about who we are and what nature is. Philosophical errors mark our exhaustion, our inability to continue. Conclusions mark the congestion and congealing of thought. We never lift the veil of Isis, but rather in our deaths yield to the silence at the heart of nature's autogenetic creative circle. Instead of a great lesson at the end of history, Schelling, too, speaks of a growing silent (*das Verstummen*), a contemplative adoration and gratitude at the heart of the study of nature. In the gravity and light essay, we are ". . . introduced into the holy Sabbath of nature, into reason, where it, resting above its transitory works, recognizes and understands itself as itself. Then it speaks to us in the measure that we ourselves fall silent" (I/2, 376).[30] At the end of the first draft of *The Ages of the World* (1811), Schelling proclaims:

> I would like to take this opportunity to say, if it were not too immodest, what I so often felt, and in an especially lively way with this present presentation: how much nearer I am than most people could probably conceive to this growing silent of knowledge [*Verstummen der Wissenschaft*] which we must necessarily encounter when we know how infinitely far everything that is personal reaches such that it is impossible actually to know anything at all. (WA, 103)

This is no mere failure, no sudden realization that silences the loquacity of Schelling's youth. This silence is the philosopher's hidden treasure, ritually enacted in the exercise of thinking.

In a sense, the problem of nature, *Naturphilosophie*, is therefore first encountered in its repression. Lore Hühn, in her marvelous book, *Kierkegaard und der deutsche Idealismus: Konstellationen des Übergangs*,[31] is right to speak of Schelling's philosophy in terms of the "return of the repressed." Deleuze, while acknowledging that there are others, writes of three discourses on blockage: the discrete, the alienated, and the repressed. In the first, infinite atoms—be they Lucretius' chance swerve by which things are revealed to be both dynamically unstable and an accident of matter or Saussure's endless supply of arbitrary but differential signifiers—are at the heart of what appears to be solid things and words. In the third, derived from Freudian psychoanalysis, repression is revealed not to be the repression of something particular that we repeat again and again. This secondary repression results from the "primary repression" of difference that produces the "representation" that we repeat again and again. It is not an issue of "brute" or "bare" repetition, that is, repression is not the condemnation to repeat what one has repressed again and again. "I do not repeat because I repress" (we do not replicate again and again a traumatic representation), but rather "I repress because I repeat, I forget because I repeat" (the primary repression of brute or bare repetition that represses difference by repeating the same) (DR, 18/29).

The second blockage, the alienation of nature, speaks most directly to Schelling's manner of approaching the issue. The concepts of nature are "concepts with indefinite comprehension but without memory" (DR, 14/24). Nature, contra Hegel's archival ferocity of orgiastic representation, does not remember to be nature. *Natura naturata* does not remember to remain itself, to repeat its identity again and again. Nature's dementia, the delirium of *natura naturans*, contests the habits of representation that maintain identities in the present.

In a sense, overcoming the alienation of nature, retrieving the living ground that modern philosophy represses (I/7, 361), is the task of what Schelling called negative philosophy, of moving through *x* to get beyond *x* (*über x hinaus*). This is the movement of *Depotenzierung*, of bringing something to its limit, of exhausting it in order to unleash what it otherwise represses. As we see in greater detail in the next chapter,

Schelling repeatedly insisted that the nature of philosophical error was not the fact that I was wrong about something, that I mistook *x* for *y*, that I failed to recognize correctly what *x* really is. Error is not a matter of holding a false position—these are trivial examples of error—but rather of allowing a position "to stand still," "to clot," and "to inhibit" the forces of its progression. We err whenever we unequivocally take a firm stand. Philosophical errancy is a question of *Hemmungen* and *Stockungen*, inhibitions and blockages. There is no final position, not even thinking's infinitely large commitment to orgiastic representation. There is no completion to thinking—it ends when we die or otherwise stop.[32] There is only the eternally new beginning of thinking and the ruse of a final position. Negative philosophy begins with clots, attempting to break through to the life that clots arrest. The error, so to speak, of modernity for Schelling lies in its inhibition of nature, but to retrieve the question of nature demands what Todd May, speaking of Deleuze's image of thought, called *palpation*:

> Concepts do not identify difference, they *palpate* it. When doctors seek to understand a lesion they cannot see, they palpate the body. They create a zone of touch where the sense of the lesion can emerge without its being directly experienced. They use their fingers to create an understanding where direct identification is impossible. . . . We might say that palpation "gives voice" to the lesion. It allows the lesion to speak: not in its own words, for it has none, but in a voice that will at least not be confused with something that it is not.[33]

But, as Deleuze argued, apropos of the three blockages, it is not enough to "suggest a purely negative explanation, an explanation by default" (DR, 15/26). How does one then speak of what one has palpated? For Schelling, this will become the problem of positive philosophy. This issue exceeds the bounds of the current chapter, but I can at least say that even before the *Freedom* essay, Schelling is already moving in its direction. In the 1806 *Aphorismen über die Naturphilosophie*, we find another remarkable claim about gravity:

> Gravity is the silent celebration of nature, in which nature celebrates unity within infinity, that is, in which nature celebrates

> its consummation. [*Die Schwere ist die stille Feier der Natur, damit sie die Einheit in der Unendlichkeit, d. h. ihre Vollendung, feiert.*] (I/7, 230)

For Schelling, gravity had a scientific, ethical, and poetic register and a properly philosophical treatment of it embraced all three dimensions of this single insight without confusing them. Gravity indicates the movement within all discrete entities by which the pretense to discretion, to independence, to being freestanding entities, was contested and brought back into a general circulation in which entities are stripped of their integrity. If one knows the Mahāyāna tradition, the force gravity attests to what is called *pratītyasamutpāda*, the dependent co-origination of things. Gravity contests what Thomas Kasulis has called the *integrity tradition of philosophy*: "In short: integrity means being able to stand alone, having a self-contained identity without dependence on, or infringement by, the outside."[34] Schelling is unequivocal about gravity's displacement of integrity: "For movement within gravity is a sign of the lack of independence of the individual, or a sign that the individual does not objectively grasp its center within itself but rather outside of itself in other things. [*Da die Bewegung in der Schwere ein Zeichen der Unselbständigkeit des Einzelnen, oder davon ist, daß es sein Centrum nicht objektiv in sich selbst begreift, sondern außer sich in andern Dingen hat*]" (I/7, 237).

In a sense, then, gravity is the counter pull, a counteracting attraction, to a thing's propensity, found within the inert force of thingliness itself, toward itself and away from a center that does not lie within itself. As such, gravity permanently threatens a thing's integrity, exposing its integrity as a lie born of inertia—for Schelling the propensity to invert good and evil, ground and existence, to make the representation of oneself the ground of all things. Inertia is contested by the counterforce of gravity. The latter is the attracting force, the force that magnetically pulls things away from themselves and into the general economy of nature. Gravity is a general center that attracts away from the inertia by which a thing strives to maintain its center of gravity within itself. Yet the depths of gravity are themselves resisted by the extensivity of light. The more the individual resists the contestation of the ground, the more it becomes full of itself, the more it swells within itself on the periphery. As Schelling describes this double contestation in *The Ages of the World*:

> But even now, intensified into selfhood (into Being-in-itself), these wholes are still retained by the attracting force. Yet, precisely because they are now selfish and because they have their own point of foundation (center of gravity) within themselves, they strive, precisely by dint of this selfhood, to evade the pressure of the attracting power. Hence they strive to distance themselves on all sides from the center of force and to become themselves away from it. Hence, the highest turgor of the whole emerges here, since each particular thing seeks to withdraw itself from the universal center and eccentrically seeks its own center of gravity or foundational point. (I/8, 323–324)

The more an individual insists on its own individuality, the more it seeks gravity within itself, the sicker it becomes. For humans, the celebration of the self in flight from the "silent celebration of nature" is the experience of evil and the possibility of such uniquely human perversity marks the very *humanity* of humanity. Resistance to gravity is the light of humanity's attempt to center itself within itself, to imagine itself as a subject for which freedom is a predicate. Freedom is what it repressed in light's obliviousness to its original darkness, in its attempt to locate its center of gravity within itself, in its delusion that it is most fundamentally attracted to itself. It is humanity's insistence on itself as an originary subject. The human, in its flight from nature itself, seeks itself as the fundamental point of reference for all predicates.

Sin is the flight from the general economy to which general gravity pulls the creaturely. It is important to note, however, that these are passive processes, not conscious choices. Although I do not chose to sin, I take responsibility for it in guilt. I affirm that in sin I have affirmed myself as my own center of gravity, holding the ground of gravity in abeyance. "The beginning of sin is when the human steps out of authentic being into non-being, out of truth into the lie, out of light into darkness in order to become a self-creating ground and to dominate all things with the power of the center that they have within themselves. [*So ist denn der Anfang der Sünde, daß der Mensch aus dem eigentlichen Sein in das Nichtsein, aus der Wahrheit in die Lüge, aus dem Licht in die Finsternis übertritt, um selbst schaffender Grund zu werden, und mit der Macht des Centri, das er in sich hat, über alle Dinge zu herrschen*]" (I/7, 391).

Yet the oblivion of gravity in the flight to the integrity of humanity does not vitiate the attracting force. Rather gravity in the ethical dimension erupts as a monstrous secret within, pulling one away from oneself in a vortex of heretofore concealed madness. God dances on volcanoes and delirium lurks within good sense. Freedom returns in the fashion that Georg Büchner's Lenz confessed to the priest Oberlin: "Do you hear nothing? Do you not hear that horrible voice that screams across the entire horizon, the one that one usually calls silence? I always hear it, it does not let me sleep."[35] Perhaps, as Schelling warns, one is "seized by dizziness on a high and precipitous summit" and "an enigmatic voice seems to cry out that one plunge from it." Or perhaps it is like that "old fable" (*The Odyssey*) in which "the irresistible song of the Sirens rings out from the depths in order to attract the passing sailors down into the whirlpool" (I/7, 382). In any case, within the cool, silent evil that is the narcissism nascent in every dogmatic self-understanding, the monstrosity of freedom can suddenly erupt, as if from nowhere, deducible from no conception of the self-represented self. This is the explosion of integrity into the orgiastic abandon of the Maenads, the frenzied reassertion of the A^2, by which Pentheus is not recognizable as a mother's son. This is the rage that Homer laments in the *Iliad*, in which the dogs of war rule and one is blinded by one's own rage, thinking oneself invincible and killing without limit. This is the "murky and wild enthusiasm that breaks out in self-mutilation or, as with the priests of the Phrygian goddess, auto-castration" (I/7, 357).

In a manner of speaking, all "natural" *Schwerkraft* is saturnalian, experienced by creatures in the same way that a seemingly self-possessed sea registers the gravitational pull from the moon, whose sway pulls the calm of the sea away from itself and tosses it beyond itself. In the tradition, of which Schelling is fully aware, the sway of Saturn also precipitated *Schwermut*, which, along with *Melancholie*, is one of the two German terms for melancholy. *Schwermut* is literally a heavy mood, a gravitational mood, the governance of the moon over all things. Zarathustra, too, sang songs of melancholy. Schelling speaks directly to the eruption of a saturnalian sensibility, of the attractive force of a virulent nihilism that pulls all things into its vortex, when he reflects in the *Freedom* essay that

> The human never receives the condition within their power, although they strive to do so in evil . . . Hence, the veil of melancholy that is spread out over all of nature, the deep, indestructible melancholy of all life. [*Der Mensch bekommt die Bedingung nie in seine Gewalt, ob er gleich im Bösen danach strebt*

> . . . *Daher der Schleier der Schwermut, der über die ganze Natur ausgebreitet ist, die tiefe unzerstörliche Melancholie alles Lebens.*] (I/7, 399)

Yet before we become monotonously gloomy, as if *Schwerkraft* could be reduced to a single tone, mood, or force, I would like to approach briefly Schelling's heavy insight from one of the great comic European novels of the twentieth century, namely, Milan Kundera's *Farewell Waltz* (1973). I do so also to open a space in which one can see that Schelling is not touting melancholy as the preeminence of sadness or to emphasize the vanity of human affairs and the futility of human thinking and action.

Kundera's novel is set largely in a fertility spa, where God, in a horribly comic revelation, appears in the sublimely ridiculous form of a Bohemian doctor who remedies cases of infertility by secretly injecting women with a syringe filled with his own semen. Dr. Skreta, whose avocation is to keep time in jazz music by playing the drums, as if he were setting and maintaining the tempo of creation itself, is the secret father of countless babies. Bohemian births are predicated on this original but concealed lie. At the end of the novel, he engages in a short conversation with his avid admirer Bertlef, who believes that not only should Jesus be celebrated because he was a womanizer ("Jesus loved women!"), but that each life has some hidden but inviolable value, a secret and inestimable value that stems from its very origin. Skreta responds by claiming, "We shouldn't be so interested in justice" because "justice is not a human thing. There's the justice of blind, cruel laws, and maybe there's also another justice, a higher justice, but that one I don't understand. I've always felt that I was living here in this world *beyond justice*."[36] Skreta, the concealed and sovereign God, immune to the call of justice, and the terrible and awful secret ground of Bohemian youth!

Meanwhile, another character, Jakub, who had tirelessly and admirably resisted the capricious despotism and fatuous dogmatism of Central European political life, has fled Bohemia because for no discernible reason whatsoever he had not told Ruzena, a nurse at the spa, that she inadvertently possessed the pill that he had always carried with him in case he required a quick exit from life. She innocently takes the little blue pill and dies. Jakub not only cannot grasp why had not told her this small but hugely consequential fact, but is unsettled by his sudden betrayal of the master values that had defined how he had lived and understood himself. Jakub muses that his deed, as senseless as it was, far exceeds the tragic experiments of Dostoyevsky's Raskolnikov:

> After Raskolnikov killed the old usurer he did not have the strength to control the tremendous storm of remorse. Whereas Jakub, who was deeply convinced that no one had the right to sacrifice the lives of others, felt no remorse at all . . . Raskolnikov experienced his crime as a tragedy, and eventually he was overwhelmed by the weight of his act. Jakub was amazed that his act was so light, so weightless, amazed that it did not overwhelm. And he wondered if this lightness was not more terrifying than the Russian character's hysterical feelings. (FW, 257)

The heaviness of gravity is not the solidity and firmness of life, and the clarity of our motives and responsibilities. This heaviness was saturnalian and, as such, paradoxically, horribly light. Such lightness emerges in the loss of one's personal center of gravity amid this deeper and more general gravity. Might not we also call such heaviness, borrowing a phrase that Kundera used later in his career, *an unbearable lightness of being*? Like the sudden urge to jump from a tremendous height? Like Edgar Allen Poe's *Imp of the Perverse*, when the narrator, who has committed a perfect crime of murder, is suddenly seized by the enigmatic motivation to start leaving telltale hints and clues?

> If there is no friendly arm to check us, or if we fail in a sudden effort to prostrate ourselves backward from the abyss, we plunge, and are destroyed. Examine these and similar actions as we will, we shall find them resulting solely from the spirit of the *Perverse*. We perpetrate them merely because we should *not*. Beyond or behind this there is no intelligible principle. . . .[37]

Jakub, for his part, as he was escaping from Bohemia, espied a small child with large glasses. In a world beyond justice, unbearably light, in which the secret ground of Bohemian youth goes by the name Skreta (yes, the world began with a bang and its awful and terrible name was Skreta), Jakub suddenly encounters the veil of melancholy spread out over all of nature:

> What fascinated Jakub about the child's face were the eyeglasses. They were large eyeglasses, probably with thick lenses. The child's head was little and the eyeglasses were big. He was wearing them like a burden. He was wearing them like a fate. He was looking through the frames of his eyeglasses as if through a wire fence. Yes, he was wearing the frames as if they were a wire fence he would have to drag along with him all his life. And Jakub looked through the wire fence of the eyeglasses at the little boy's eyes, and he suddenly was filled with great sadness.
>
> It was as sudden as the spread of water over countryside when riverbanks give way. It had been a long time since Jakub had been sad. Many years. He had known only sourness, bitterness, but not sadness. And now it had assailed him, and he could not move.
>
> He saw in front of him the child dressed in his wire fence, and he pitied that child and his whole country, reflecting that he had loved this country little and badly, and he was sad because of that bad and failed love. (FW, 270–271)

Jakub, who had not consciously chosen hate, suddenly took responsibility for it, for having been lost in himself, in his self-righteous politics, and all of the other symptoms of his "bad and failed love." He was suddenly afforded an insight into his sickness precisely in glimpsing and taking responsibility for a failure that was unthinkable within the clarity of his moral self-understanding. Yet it would be wrong to read this as the monotonous triumph of the melancholic mood. Let us also remember the surreptitious delight of reading Poe and the brilliant comedy of both *Farewell Waltz* and Zarathustra's braying at the Ass Festival. As Schelling claimed in the *Freedom* essay, "Joy must have suffering in order to be transfigured into joy [*Freude muß Leid haben, Leid in Freude verklärt werden*]" (I/7, 399). *Leid*, the word for suffering, also carries the connotation of passivity, of being pulled away from itself and being formed into something else, of the free activation of *der ewige Anfang*, the eternal beginning, of a sea under the silent governance of the moon. The dark night is also a sun. In the *Stuttgart Lectures*, Schelling called light "positive darkness—evolution."[38] "All birth is birth out of darkness and into light" (I/7, 360–361) and that consequently the "birth of light is the realm of nature" (I/7, 377–378). Nature, the splendor of light, is also rife with the profundity of its multiplicity. Light is *natura naturans*, nature naturing—in no way to be confused with natural objects. It is rather nature's unbidden mode of appearing, its very coming to light, so to speak.

You Are Not What You Were But as Though Emerging from Air, You Slowly Show Yourself as Someone, Not Ever Remembered

We move now from the passage of "heres" to the passage of "nows," that is, from the movement of space to the movement of time. Presumably critiquing Habermas, Deleuze and Guattari in *What is Philosophy?* do not seek the conditions of a rejuvenated philosophy in "the present form of the democratic State or in a cogito of communication." They discern no dearth of communication. "On the contrary, we have too much of it. We lack creation. *We lack resistance to the present.* The creation of concepts in itself calls for a future form, for a new earth and people who do not yet exist" (WP, 108).

When Nietzsche extolled Schopenhauer as an educator, he did not do so because he agreed with Schopenhauer's actual thinking, but rather because Schopenhauer went his own way and was not cowed by convention. He was *unzeitgemäß*, counter to the measure of his present. Nietzsche recalls that a world traveler had reported that people of all lands have a "propensity for laziness [*Hang zur Faulheit*]," itself, Nietzsche surmised, the symptom of "fear of the neighbor, which convention demands and with which one veils oneself."[39] Education then becomes the cowardly indolence of the *herdenmäßig* (the measure of the herd), the bare repetition of the present.

Deleuze writes of three passive syntheses in which the temporality of repetition plays out. In the first, that of habit, presence is the rule of the day in the *herdenmäßig* repetition of the habits that make possible the recognizable, the obvious, and the familiar. Yet, habit, as Bergson demonstrated, is the art of bringing the past forth into the present, of finding adaptations to the otherwise unwieldiness and unviability of matter. In the second passive synthesis, that of memory, we see that Proust was right: the archive of memory is comprised of virtual objects that are transformed in being brought forth into the present. The past always has a future, that it is not the fixation of habits of recognition and the tyranny of good sense, but the creativity and transformation *rechercher*—of calling for the past in new ways. This reveals the third passive synthesis, the synthesis lurking in the depths of habits whose techniques of selection create the appearance of representation. This is the "royal repetition" in which the "ground has been superseded by a groundlessness, a universal *ungrounding* [*effondement*] which turns upon itself and causes only the yet-to-come [*l'à-venir*] to return" (DR, 91/123). The future ruptures the dogmatic image of thought, and wrecks havoc with good sense.[40] The rift that the future exposes is what Hölderlin famously called the *caesura*, "as a result of which beginning and end no longer coincided. We may define the order of time as this purely formal distribution of the unequal in the function of the caesura" (DR, 89/120).

But the infinite distance of the future is not for Schelling lost in the infinite depths of nature. In a striking fashion, *der ewige Anfang* is already a premonition of the eternal return of the same, and this shall comprise the theme of our final meditation.

Iain Hamilton Grant, in his seminal *Philosophies of Nature after Schelling*,[41] has impressively indicated the manner in which Schelling, as well as the comparative anatomist Carl Friedrich Kielmeyer, broke free from the trap of linear recapitulation. The latter asserts, as classically stated in the now largely refuted Meckel-Serres Law, that ontogeny recapitulates phylogeny. The somatic development of the individual, that is, the developmental stages that govern the shape and size of a given organism, repeats again the laws that govern the evolutionary development of that organism's species. In Meckel's formulation, "the development of the individual organism obeys the same laws as the development of the whole animals series" (PON, 129).

Such development cannot be dynamically genetic or truly progressive, because the recapitulation that governs the series repeats itself without deviation. The RE indicates that it capitulates, that is, reinstantiates the main rules that govern the growth of the organism, without those rules being displaced or otherwise interrupted. Every microcosm is in each case a summary of its macrocosm, and hence the history of a species is continuous, that is, a continuation of the same developmental rules. Kielmeyer countered this by deriving variation not from the homogenesis of bodies, but rather the "heterogenesis" (PON, 120) of forces and powers (*Potenzen*) (PON, 122), prompting Schelling to announce that Kielmeyer had inaugurated "the advent of a completely new epoch in natural history" (PON, 120).

What makes possible the progressive stages of becoming? That is, how can one release the long-term nonlinear recapitulation repressed within our short-term sense that recapitulation is linear? Already in the nineteenth century, natural historians had to grapple not only with Darwin, but also just as dramatically with the shock of the increasing awareness of a fossil record that indicated ruinous squandering, a delirium of speciation. Long, short, history: these are all figures of time and it is the latter that interrupts the would-be linearity of recapitulation. *Time reveals recapitulation to be the repetition of difference, not the recursion of representation.* "Repetition is a transgression. It puts law into question, it denounces its nominal or general character in favor of a more profound and more artistic reality" (DR, 3/9).[42] Nonlinear recapitulation is genetically dynamic because it is not the recursion of the self-same procedures through time, but rather the progressive force of time itself. This is not the history of formal laws determining the forms of bodies recurring in time, such that bodies are always recognizable in their enduring representations. In one of Schelling's formulations, nature is auto-poietic, creating itself out of itself, "self-recapitulating at different levels" (I/4, 47; PON, 174).

When time is not linear, it is, as Hamlet lamented, "out of joint." Time is the discontinuous play of forces in the taking place of space, not the causal relations of discrete bodies in space. As Deleuze puts it: "In other words, reality is not the result of the laws which govern it, and a saturnine God devours at one end what he has made at the other, legislating against his creation because he has created against his legislation" (DR, 227/292–293). Time, *natura naturans*, shatters the somatism of modern philosophy and science, revealing the productivity of forces at the heart of rules and forms.[43] In the words of Benjamin Paul Blood (1822–1919), whom Deleuze approvingly cites: "Nature is miracle all. She knows no laws; the same returns not, save to bring the different" (DR, 57/81). The disparity that repeats itself ever anew as the inexhaustible "clamor of being"[44] was already intimated by the young Schelling and reconfirmed by Kielmeyer. With Grant:

> Kielmeyer nowhere asserts "that the series of beings is linear, so that the ontogeny of man recapitulates the phylogeny of the entire animal kingdom." In place of the somatic poles of mega- and micro- bodies—protoplasm, fetus, man and world, substrate and atom, species and individual, substance and concrete

> whole, type and phenomenon, and so forth—by which the *linear usage of recapitulation* is maintained, the "object" of Kielmeyer's *non-linear use recapitulation* is *time* . . . Kielmeyer's "series of organizations" are not the *object* of natural historical explanation, but the *medium* through which time becomes momentarily phenomenal. (PON, 134)

Recapitulation is not the eternal return of either the same rules or their recursion in the same phenomena. Neither the same rule nor the same event repeats itself as itself infinitely. What returns is not the *object* of the return, but rather the *medium* of the return, that is, the *return itself, die Wiederkehr selbst*, eternally returns as the discontinuity and disequilibrium of time in its progression as the becoming of *natura naturans*. Deleuze argued therefore that the eternal return of the same is the "principle" of "difference and its repetition" (NP, 49/55), and, as such, it is "an answer to the problem of *passage*." If beings are, as in Zeno's famous paradoxes, themselves in time, how can the same being remain itself yet move through time? If it moves through time, it cannot be the same being in $time_1$ and $time_2$; it is itself as the being at rest with itself; if it is itself, it is at rest as itself in a single point in time.

Bergson famously demonstrated the illusion of this would-be paradox: "they all coincide in making time and movement coincide with the line which underlies them."[45] Passage is misunderstood as points in space trying to move through time while remaining themselves. Time reveals that they were not themselves. "It is a law of our representation that the stable drives away the unstable, the important and central element for us becomes the atom, between the successive positions of which movement then becomes a mere link." The deflation of the problem of matter to atomic representations "is hardly anything but an outward projection of human needs" (MM, 203). Moreover, it is also a "fiction" that there is homogenous time, recursively maintaining its identity as the abstract "flowing of time." There is no "homogenous and impersonal duration, the same for everything and everyone, which flows onward, indifferent and void, external to all that endures." For Bergson, there is no master rhythm of time, no fixed rule of time, no phylogenesis of time governing the ontogeny of lived times. It is "possible to imagine many different rhythms which, slower or faster, measure the degree of tension or relaxation of different kinds of consciousness and thereby fix their respective places in the scale of being" (MM,

207). Perception therefore "means to immobilize," that is, to condense "enormous periods of an infinitely diluted existence into a few more differentiated moments of an intenser life." We "seize" in the "act of perception . . . something which outruns perception itself" (MM, 208).

Although Bergson is speaking of memory's capacity to enable the perception of matter as various durations, this relationship between time and becoming, beyond the "exigencies of life" (MM, 208) and the adaptations that they require, was the heart of the question of nature for both Schelling and Deleuze's reading of Nietzsche.

Time is the gravitational counterforce to rest, the animating and animated mobility and hence variability of matter. In the 1806 addendum to the *Weltseele* on the oppositional relationship of gravity and light, Schelling claims "what concatenates this contradiction into itself is time [*das Verknüpfende dieses Widerspruchs in ihm selbst aber ist die Zeit*]" (I/2, 365). Gravity reveals itself as light, while remaining withdrawn from the light precisely as it withdraws the light away from itself, disallowing it from coming to a rest in itself. This is the animating movement of *das Band*, the one that can only show itself as many without in itself being anything and without allowing anything to complete itself in itself. Gravity, "in itself eternal" (what has always in itself not happened in anything that happens), nonetheless also reveals itself, comes to light, that is, takes place as time, as the univocal clamor of being, as "time in the bonded as the bonded [*ist in dem Verbundenen, als Verbundenen, die Zeit*]." Eternity is *zeitlich gesetzt* (posited temporarily), and the monstrous gravitational expulsion and repulsion of light is the animating imagination of the whole into the individual ("*Beseelung ist Einbildung des Ganzen in ein Einzelnes*") (I/2, 364). Although the problem of the imagination comprises the final chapter of this book, we can already appreciate that it names the unprethinkable creativity of time.

In *The System of Transcendental Reason*, "succession and all changes in time are nothing else but the evolutions of the absolute synthesis" (STI, 157). Schelling calls this the *schematism of time* and argues that it addresses the famous problem of passage, taking up the problem almost a century before Bergson:

> In every alteration a transition takes place from one state to one that is contradictorily opposed to it, for example, when a body transitions from movement in direction A to movement in direction ¬A. . . . Intuition produces time as constantly in

> transition from A into ¬A in order to mediate [*vermitteln*] the contradiction between opposites. Through abstraction, the schematism and with it time are sublimated [*aufgehoben*].—There is a well known sophism of the ancient Sophists whereby they contest the possibility of communicating motion. Take, they say, the last moment of rest of a body and the first moment of its movement. There is no intermediary between the two. (This is also completely true from the standpoint of reflection.) So if therefore a body should be put in motion, then this happens either in the last moment of its rest or in the first moment of its movement; but the former is not possible because the body still rests and the latter is not possible because it is already in movement. (STI, 188/144)

The schematism of time, mediating inner and outer sense, is not the disconnected series of moments of being (e.g., A is resting, something other than A is moving. If A becomes ¬A, that is, transitions from rest to movement, either it was not itself, that is, not a resting A, or, if it is a resting A, then it cannot become a moving A). The problem of passage was originally putatively solved by "artifices of mechanics," but Schelling's solution is more far reaching. It is not to find a mechanical intermediary pushing A from itself to something else, or imagining that A is somehow identical to itself, but modified by accidents like rest and motion. Infinity schematizes or imagines itself in each moment anew: ∞/A, ∞/B, ∞/C.

Not even science escapes the problem of passage. There is nothing that in itself is absolutely true. As we read in the *Weltalter*:

> But to speak the truth, it is no less the case with true science than it is with history that there are no authentic propositions, that is, assertions that would have a value or an unlimited and universal validity in and for themselves or apart from the movement through which they are produced. Movement is what is essential to knowledge. When this element of life is withdrawn, propositions die like fruit removed from the tree of life. Absolute propositions, that is, those that are once and for all valid, conflict with the nature of true knowledge which involves progression. Let, then, the object of knowledge be A and then the first proposition that is asserted would be that

> "A = x is the case." Now if this is unconditionally valid, i.e., that "A is always and exclusively only x," then the investigation is finished. There is nothing further to add to it. But as certainly as the investigation is a progressive kind, it is certain that "A = x" is only proposition with a limited validity. It may be valid in the beginning, but as the investigation advances, in turns out that "A is not simply x." It is also y, and it is therefore "x + y." One errs here when one does not have a concept of a kind of true science. They take the first proposition, "A = x," as absolute and then they perhaps get, or have in mind from somewhere else in experience, that it would be the case that "A = y." Then they immediately oppose the second proposition to the first instead of waiting until the incompleteness of the first proposition would demand, from itself, the advance to the second proposition. For they want to conceive of everything in one proposition, and so they must only grant nothing short of an absolute thesis and, in so doing, sacrifice science. For where there is no succession, there is no science. (I/8, 208–209)

This was also how Schelling tried to breathe life back into and reanimate what he feared was a dangerously dogmatic quality of Spinoza's substance (a "one-sided realism" he says in the *Freedom* essay). There is no tautology between God and things (God=things; ideal=the real) or eternity and time. God and things are not *einerlei*, that is to say, they are not one of a kind. Hence, even a simple judgment like "*x* is *y*" does not say that "*x*=*y*," but that the copula joins together or synthesizes *x* and *y*, despite their distinctness. The judgment, "the body is blue" does not assume that the idea of the body includes the idea of blueness or vice versa. The copula synthesizes them while preserving their difference (I/7, 341). Hence if one were to designate the Absolute as A and the modes of its attributes as A/a, it does not follow that A = A/a (I/7, 344). Schelling links A/a, A/b, A/c to what Leibniz called monads (I/7, 344–345), which, in the present context, we can also call *moments*.

In the *System of Transcendental Idealism*, Schelling already claimed that each moment of time is already the whole of time: "what one dubs various times are only the various delimitations [*Einschränkungen*] of absolute time" (STI, 189; 145). Present moments are not isolated atoms of time, but rather the progression ∞/A, ∞/B, ∞/C is in each moment that ∞ displaces itself into the past in order to present itself precisely as present, but in so doing, the disequilibrium between the past ground and presence (the literal meaning of *Dasein*, existence, to be there) haunts the present as the future, the darkness of the *Ungrund* as the intimation of the future vitally indwelling in what is there.

Schelling's most thoroughgoing and profound discussion of time is found in the first or 1811 draft of *Die Weltalter*. Here Schelling distances himself from both Kant's reduction of time to an *a priori* form of representations (*Vorstellungen*) and his insistence on the regulative subsistence of a single subject:

> No thing has an external time. Rather, each thing only has an inner time of its own, inborn and indwelling within it. The mistake of Kantianism with respect to time consists in it not knowing this *universal* subjectivity of time and hence it delimits time in such a way that it becomes a mere form of our representations [*Vorstellungen*]. No thing comes to being in time. Rather *in* each thing time comes to being anew and does so immediately from out of eternity. (WA, 78–79)

This was also Schopenhauer's mistake: that the transcendental causality of time and space is one thing. As Deleuze felicitously frames the problem:

> And Nietzsche's break with Schopenhauer rests on one precise point; it is a matter of knowing whether the will is unitary or multiple. Everything else flows from this. Indeed, if Schopenhauer is led to deny the will it is primarily because he believes in the unity of willing. Because the will, according to Schopenhauer, is essentially unitary, the executioner comes to understand that he is one with his own victim. The consciousness of the identity of the will in all of its manifestations leads the will to deny itself, to suppress itself in pity, morality and asceticism. Nietzsche discovers what seems to him the authentically Schopenhauerian mystification; when we posit the unity, the identity, of the will we must necessarily repudiate the will itself. (NP, 7)

When Schelling speaks of *das Urlebendige* or primordial life as One, and as having "everything within it at the same time" and being "without distinctions [*in ihm liegt alles ohne Unterscheidung zumal, als Eins*]" (I/8, 201), this One is not mathematically one; it is not *einerlei*, one thing. It is not a comprehensible first principle or a discernible rule lying at the heart of things. In itself it is unruly, dark, absolutely silent, Plato's χώρα. Time is not therefore an a priori law. It is not a generality, governing all instances of appearance. And in a powerful re-articulation of the temporality of nonlinear recapitulation, Schelling goes on to claim "every single individual comes to being [*entsteht*] through the same cision (*Scheidung*] through which the world comes to being, and therefore does so right from the beginning with its own epicenter of time" (WA, 79). Individuals emerge from the monstrously generative disparity, the dis-equal cut or gap, *die Scheidung*, between the *Ungrund* of eternity and the being of becoming.[46] In this sense, one cannot speak of space and time as the recursive rules of intuition and representation in the linear development of experience. Space is not a series of atomistic points somehow passing through time. If time expresses itself as the being of becoming, as the spatial extension of time's intensity, it does not follow that time is the rule or rules that govern its extension. Time is not homogeneous.

Events are not the linear recapitulation of the rule of time. As Schelling put it the *Weltalter*: "nature reserved for itself to renew constantly each moment in the present time" (I/8, 290). In this sense, time and space are not the pure forms of intuition. As Deleuze formulates it: "The reason of the sensible, the condition of that which appears, is not space and time but the Unequal in itself [*l'Inégal en soi*]" (DR, 222–223/287). Deleuze names the Unequal in itself *la disparité*, intensity as difference (DR, 222/287).[47] As the sufficient reason of all phenomena, it is exhausted by none of its appearances, but is rather as "difference without a concept, non-mediated difference" (DR, 25).

Rather than the mechanical repetition of linear recapitulation, "cyclical generalities in nature are the masks of a singularity which appears through their interferences; and beneath the generalities of habit in moral life we rediscover singular processes of learning. The domain of laws must be understood, but always on the basis of a nature or a Spirit superior to their own laws, which weave their repetitions in the depths of the earth and of the heart, where laws do not yet exist" (DR, 25/38). The representation of nature, the aspiration of modern philosophy and science, deprives it, as Schelling charged in the *Freedom* essay, of its living ground (I/7, 361).[48]

Representation assumes sameness, whether identities or known in themselves or through their forms of representation. For Schelling, on the other hand, the absolute is neither a noumenal thing in itself nor a phenomenal thing. (That it is not yet a thing is the literal meaning of *unbedingt.*) In *The System of Transcendental Idealism* (1800), Schelling argues that the pure *Ich*, whose multiplicity cannot be reduced to any one thing in particular, "is an act that lies outside of all time" and hence the question "if the I is a thing in itself or an appearance [*Erscheinung*]" is "intrinsically absurd" because "it is not a thing in any way, neither a thing in itself nor an appearance" (I/3, 375). It is literally *unbedingt*, that which has not and cannot become a thing within each thing. As such it is sovereign and hence unequal to its either its appearances or to the rules that would fix those appearances.

In this way, Schelling is clear that the question "if the world has always existed from infinite time or if it has existed since a particular time" (WA, 79) is not well posed. The concept of "infinite time" is "absurd," but the second possibility, that it "has existed since a particular time" is also highly problematic. At what time did time begin? When did the world come into time? These are questions that assume time in order to explain time: there was a time in which time came to be. "Each beginning of time presupposes a time that has already been" (WA, 79) and hence "a beginning of time can in no way whatsoever be thought" (WA, 79). We can never ask, "Since when did time begin?" or "How long has time already lasted?" (WA, 79).

Rather, time "in each moment is the whole of time, that is, the past, present, and future, that begins not in the past, not from the limit, but rather from the epicenter and in each moment is equal to eternity" (WA, 80). From this epicenter, itself opposed to the "common mechanical concept of time" and the "illusion of abstract time [*Scheinbild einer abstrakten Zeit*]" (WA, 79), one asks: "How many times have their already been [*Wie viele Zeiten sind schon gewesen*]?" (WA, 80). "Every particular time already assumes time as a whole" (WA, 81), the temporal animation of matter, and hence one can also say that time is organic in the whole and hence also organic in each particular thing. Without such organisms, "all of history would be a chaos full of incomprehensibilities" rather than what it is, namely, "temporal unities [*Zeiteinheiten*]" that as "periods . . . present in themselves the whole of time" (WA, 82).

Schelling then asks: "But what is the organizing principle of these periods?" Clearly it is not any form of time, for time itself is the very process of imagining coming into a single image (*Einbildungskraft*, *Ineinsbildung*). "Without a doubt," Schelling argues, it would be what "contains time as a whole." Schelling continues, evoking a Trinitarian mythology of time, with time, the A^3, the spirit that expresses the unity of the eternally and gravitationally dark God and the expansive light of the Son (the realm of the image):

> But the whole of time is the future. Therefore, only the spirit [*Geist*] is the organic principle of times. The spirit is free from the antithesis of the contracting force of the Father and the expanding force of the Son. In the Spirit, first of all, both again reach consummate equality . . . When the force of the Father is posited in the past in relationship to the Son, it is no way the view that it is posited as not having being at all [*als überall nicht seyend*]. This force is posited as not having the being of the present but nonetheless as having being and acting in the past . . . But the will of the Father in relationship to the Son and of the Son in relationship to the Father is the will of the spirit. The spirit knows in which measure [*Maß*] the eternal concealment [*Verborgenheit*] should be opened and posited as past . . . Only the spirit explores everything, even the depths of the Godhead. The science of the coming things rests in the spirit alone. Only the spirit is entitled to loosen the seal under which the future lies decided. For this reason the prophets are driven by the spirit of God, because this alone is the opener of times: for a prophet is the one who sees into the belonging together of the times. (WA, 82–83)

As we saw in the second chapter, this is the problem of the future and of prophecy, which demands that we think the unity of time as such. In the 1815 *Weltalter*, God as the NO (contracted gravity) and the YES (the SON as light) are again thought together in their unity:

> The contradictory relationship is resolved through the relationship of the ground by which God has being as the No and the Yes, but one of them as prior, as ground, and the other as posterior, as grounded.

> As such, it always remains that if one of them has being, then the other cannot have the *same* being. That is, it remains that both exclude each other with respect to time, or that God as the Yes and God as the No cannot have being *at the same time*. We express it intentionally in this way for the relationship cannot be of the kind such that if the posterior, say A, has being, then the posterior, hence B, would be sublimated, or simply ceased to have being. Rather, it always and necessarily abides as having the being *of its time*. If A is posited, then B must just still persist *as the prior*, and hence, in such a way, that they are nonetheless, at the same time, *in different times*. For different times (a concept that, like many others, has gotten lost in modern philosophy) can certainly be, as different, at the same time, nay, to speak more accurately, they are necessarily at the same time. Past time is not sublimated time. What has past certainly cannot be as something present, but it must be as something past at the same time with the present. What is future is certainly not something that has being now, but it is a future being at the same time with the present. And it is equally inconsistent to think of past being, as well as future being, as utterly without being. (I/8, 301–302)

This is the sovereign and utterly free Godhead at the heart of time, its dark past and its ongoing presentiments of the unprethinkable:

> The Godhead is nothing because nothing can come towards it in a way distinct from its being [*Wesen*] and, again, it is above all nothingness because it itself is everything.
>
> It certainly is nothing, but in the way that pure freedom is nothing. It is like the will that wills nothing, that desires no object, for which all things are equal and is therefore moved by none of them. Such a will is nothing and everything. It is nothing in so far as it neither desires to become actual itself nor wants any kind of actuality. It is everything because only from it as eternal freedom comes all force and because it has all things under it, rules everything, and is ruled by nothing. (I/8, 235)

In this sense, I concur here with two important points that Grant has made in this respect. First, as Schelling insists already in 1797 (*Ideas for a Philosophy of Nature*), philosophy is the "infinite science"

(I/2, 256). Philosophy does not eliminate "anything a priori from its remit" (PON, 19–20). There is no philosophical problem or theme that philosophy dismisses in advance. The future of thinking, including the future thinking regarding what has already been thought, remains *unvordenklich* and the miracle of its natalities cannot be circumscribed in advance. It follows that *Naturphilosophie* is inexhaustible—an image of thought, not a philosophical agenda, the infinite exercise of philosophizing and not this or that philosophy. Second, Grant is spot on to claim that Schelling's "identity differentiates rather than integrates" and that "unlike Kant's transcendentalism, there is no single subject at the core of Schelling's philosophy, but rather a multiplicity of subjects" (PON, 174). Indifference is not one nor does it lack individuality and singularity because time is not homogeneous.

In this respect, Deleuze perhaps should not have distanced Schelling from Spinoza and Nietzsche when he charged that the *Ungrund*, which is the miraculous progressivity of *der ewige Anfang* after all, "cannot sustain difference" (DR, 276/354). When Schelling claimed in 1797 that "Nature shall make the spirit visible, and the spirit, nature invisible" (I/2, 56), that is, the "absolute identity of Spirit *in us* and nature *outside of us*," this is also implicitly to have said that in *Naturphilosophie*, that is, in philosophizing the way that nature natures, philosophy in its own way expresses nature's infinite, differential prodigality.

Finally, when "you are not what you were but as though emerging from air, you slowly show yourself as someone, not ever remembered," this is in a certain respect to see the proximity of Schelling's potencies in the miracle of their ongoing natality to Deleuze's interpretation of Nietzsche's *ewige Wiederkehr des Gleichen*. It is not the same thing that returns eternally, but rather the eternal returning of the return itself. The return itself always returns, not any particular object of the return.[49] This, again, is the problem of passage, beyond the good sense that generates Zeno's paradoxes. It is the eternal repetition of difference:[50]

> It is not being that returns but rather the returning itself that constitutes being insofar as it is affirmed of becoming and of that which passes. It is not some one thing which returns but rather returning itself is the one thing which is affirmed of diversity or multiplicity. In other words, identity in the eternal return does not describe the Nature of that which returns but, on the contrary, the fact of returning for that which differs. (NP, 48/55)

Le revenir lui-même, the coming that always comes again, the implacable advent of the RE itself, the turning that ever turns anew, is *der ewige Anfang* in the infinite depth of its past and the infinite distance of its *unvorkenliche* futurity. "It is in difference that phenomena fulgurate and express themselves as signs (*fulgure, s'explique comme signe*), and movement is produced as an 'effect'" (DR, 57/80, translation altered).[51] Nietzsche called thinking's attainment of this image of thought *affirmation*, the holy Yes-saying. Schelling called it *love*. In the *Freedom* essay, love emerges as the concession (*Zulassung*) to ground, in *das Wirkenlassen des Grundes*, in letting the ground operate (I/7, 375). Love is a "unity that is the same towards everything, but clasped by nothing [*gegen alles gleiche und doch von nichts ergriffene Einheit*] . . . it is the all in all [*das Alles in Allem*]" (I/7, 409). One could say that love is the affirmation of the solitude of God. Love loves the personality of Nature as the progressive recapitulation of ground, that is, the repetition of difference. Nature? What are you now? Well then. Once more! *You are not what you were but as though emerging from air, you slowly show yourself as someone, not ever remembered.*

4

Stupidity

> In the preface to the *Phenomenology*, Hegel chastised Schelling for placing stupidity at the origin of being. Hegel, for once, was unnerved. Clearly, the imputation of originary stupidity to human *Dasein* was an "issue" for Hegel, tripping him up, effecting a phenomenal misreading. Schelling posits a primitive, permanent chaos, an absence of intelligence that gives rise to intelligence. Presumptuous man has refused to admit the possibility of such abyssal origins and is seen defending himself with moral reason.
>
> —Avital Ronell, *Stupidity*[1]

The Transcendental Question of Stupidity, or the Dogmatic Image of Thought

In *Difference and Repetition*, Deleuze asks us to take up the problem of stupidity as a properly transcendental question: "how is *bêtise* (and not error) possible?" (DR, 151/197). This chapter proposes to make a contribution to this transcendental problem and to suggest that at the heart of Schelling's philosophical project is not a particular philosophy (a fixed position or an unswerving philosophical commitment), but rather an imageless image—itself nonrational and meaningless—of thought targeted at the immense stupidity and incipient madness of the dogmatic image of thought. Stupidity has nothing significant to do with a lack of intelligence. Quite the opposite: the disease of stupidity in its philosophical expression can be seen with an almost sui generis clarity in the work of "intelligent" philosophers.

Schelling spoke of a delirium at the heart of all things, "the turning wheel of birth, that wild self-lacerating delirium that is still now the innermost of all things" (WA, 43), and in Ronell's perspicacious account, this "unnerved" Hegel. Perhaps we might say that it drove him to the periphery and drove him to his own kind of transcendental stupidity, despite the paradoxically enormous brilliance of his work. Here we get our first glimpse of the problem: the depths of life drive us to the periphery where thinking, anxious about its annihilation, stupidly sublimates the center into itself. To do otherwise is madness, but then again, for Schelling, philosophy is nothing less than the capacity to negotiate this madness and navigate its errancy. "For in what does the intellect prove itself than in the coping with and governance and regulation of madness?" (I/8, 338). Before its intuition of the infinite, philosophy is called to create and deploy concepts that unleash, not sublimate, the monstrous creativity of the original intuition. Deleuze and Guattari called such philosophical concepts a *chaoid* (*chaoïde*), an ordering of primordial chaos, "a chaos rendered consistent, become Thought, mental chaosmos. And what would thinking be if it did not constantly confront chaos? Reason shows us its true face only when it 'thunders in its crater [*la raison tonne en son cratère*]'" (WP, 208/196). The last phrase, from the famous communist anthem *L'internationale*, already suggests the revolutionary possibilities of a less stupid humanity and a less stupid relationship to the earth.

Nonetheless, in the rare moments when we are honest with ourselves, if such moments ever chance to transpire, what other kinds of words evoke more anxiety and incite, or perhaps unleash, more violence, than a phrase like: *Mutter, ich bin dumm*, *Maman, je suis bête*, Mother, I am stupid? Although I will attempt not to conflate carelessly *die Dummheit*, *la bêtise*, and stupidity, I would like to follow a thread that can be first found in Schelling's 1809 *Freedom* essay and trace it primarily through the thinking of Deleuze and Flaubert by way of Nietzsche and Musil. *How does stupidity expose the violence and madness incipient but repressed within dogmatic thinking?*

In the opening monologue of Béla Tarr's last film, *The Turin Horse* (2011), a cinematic meditation on what happened to the horse that Nietzsche embraced that fateful January morning in 1889, we hear that:

> In Turin on 3rd January, 1889, Friedrich Nietzsche steps out of the doorway of number six, Via Carlo Alberto. Not far from

> him, the driver of a hansom cab is having trouble with a stubborn horse. Despite all his urging, the horse refuses to move, whereupon the driver loses his patience and takes his whip to it. Nietzsche comes up to the throng and puts an end to the brutal scene, throwing his arms around the horse's neck, sobbing. His landlord takes him home, he lies motionless and silent for two days on a divan until he mutters the obligatory last words, "*Mutter, ich bin dumm*," and lives for another ten years, silent and demented, cared for by his mother and sister.

Far worse than death for a thinker as brilliant as Nietzsche: *Mutter, ich bin dumm. Maman, je suis bête.* Mother, I am stupid. Nietzsche, the prophet of the *Übermensch*, is now utterly infantilized, struggling to understand even trivial ideas. As Rustam Singh poignantly articulates it: "For twelve long years Nietzsche could not think. He could neither think nor remember, nor remember to think, remember to go back to his thought, the only thing that he would have liked to remember."[2] The disaster had struck and Nietzsche was numb. The source of Tarr's anecdote is not clear to me. Although I do not want to deny him the use of poetic license, I am guessing that it comes from an unsettling report recorded in 1893 by Heinrich Lec, who collected some anecdotes from Nietzsche's mother, Franziska, while her son was under her care in Naumburg. The report, which appeared in a newspaper (the *Berliner Tageblatt*), itself no goldmine of wisdom, is rife with Franziska's insipid observations: she makes her son stick to a diet—"milk and honey, he likes to eat that"—, the house where Nietzsche now stays is "off the beaten track," she recites him poems that she knows that he cannot understand, and so on. Franziska also recounts the following, quite striking conversation that perhaps inspired Tarr's monologue:

> "My mother," he said to me, "I am not stupid."
>
> "No, my dear son," I say to him, "you are not stupid, your books are now world shaking."
>
> "No, my mother, I am stupid."[3]

This is a rich interchange. Franziska's incessant clichés, including, as the *pièce de résistance*, that her son is not stupid because his books are "now world shaking," are in their own way quite stupid. So what if the books are world shaking? Has not our poor world been shaken again and again by stupid thoughts? To assume that a book is a good book because everyone reads it commits an *ad populum* fallacy at the level of taste and it does not address the more fundamental value question: is the book *worth* reading? At this point we come to the first paradoxical feature of stupidity: If you know that you are stupid, you are not so stupid, because you are wise enough to know that you are stupid. The stupid, on the other hand, are too stupid to know that they are stupid. Full throttle stupidity is impenetrable. The books that Franziska could never comprehend, let alone affirm, are celebrated as world shaking.

The irony is completely lost on her when she recounts that her son told her, upon the appearance of a new edition of one of his works, "My mother, don't read it, that can't give you any joy" (CN, 231). This is true. Although she is clearly not the sharpest knife in the drawer, I suspect that she is intelligent enough to understand that she has no idea what her son's books are about and that therefore she is in no position to offer an assessment of them. The fact that she esteems them nonetheless has little to do with her meager cognitive capacity. Stupidity should not be confused with being dimwitted. Her assessment is not a mistake, but rather something much more difficult to ferret out: a commitment to judgments and convictions whose veracity or lack thereof is not at all an issue for her. In this vein, Franziska soon recounted that the "preacher from the cathedral came to see me. No one folded his hands like Fritz, he said, so sincerely and piously and devoutly. All our friends called him: the little pastor" (CN, 231). Nietzsche, the little pastor!

On the other hand, Nietzsche is right: *Meine mutter, ich bin dumm.* In his madness he is somehow not so stupid because he has some glimpse that he has collapsed into an immense, relentless, and punishing stupidity. This is already part of the poignancy that Béla Tarr's opening monologue evokes. Although Nietzsche had at best an ambiguous relationship to Socrates and his legacy, he nonetheless praised these Greek philosophers for taking on the happy and good conscience of those who take refuge in "thoughtlessness and stupidity [*Gedanklosigkeit und Dummheit*]." "These philosophers did harm to stupidity" (*Gay Science*, aphorism 328). Although Nietzsche will not here decide if this "sermon" against stupidity has better reasons to support it than those that gave rise to the alleged selflessness that bolsters and reinforces "the herd instincts," it is clear that the Greek refusal of the happy-go-lucky character of stupidity hints at an alternative to the illusion that ignorance is bliss and that joy is found in the selfless devotion to group think. It is worth noting, however, that although Nietzsche clearly embraces the Socratic refusal of stupidity's good conscience, he is ambivalent about philosophy's capacity to provide a better alternative ("We will not here decide if this sermon against stupidity had better reasons for it than did the sermon against self-obsession [*Selbstsucht*]"). If Nietzsche is against stupidity, even though Greek philosophy's argument against it is inadequate, does this not suggest that there can be ironically *stupid arguments against stupidity*? Such arguments would not be unintelligent arguments against stupidity, but rather the paradoxical stupidity of a certain kind of intelligence, what the great Austrian novelist and philosopher Robert Musil called "intelligent stupidity."[4] The fundamental problem is not that a lack of intelligence makes us prone to errors. We can, rather think perfectly true things that are utterly stupid, but not therefore somehow false.

Deleuze eloquently argues this point in *Nietzsche and Philosophy* (1962) in a chapter appropriately called "The New Image of Thought":

> Stupidity [*la bêtise*] is not error or a tissue of errors. There are imbecile thoughts, imbecile discourses, that are made up entirely of truths; but these truths are base, they are those of a base, heavy and laden soul. The state of mind dominated by reactive forces, *by right*, expresses *stupidity* [*la bêtise*] *and, more profoundly, that which it is a symptom of: a base way of thinking*. In truth, as in error, stupid thought [*la pensée stupide*] only discovers the most base—base errors and base truths that translate the triumph of the slave, the reign of petty values or the power of an established order.[5]

As we can see from Deleuze's lexicon in this passage, he is not, despite their lack of metaphorical equivalence, proposing as a large a wedge between *la bêtise* and stupidity (*la pensée stupide*) as Derrida would later suggest. There is a univocity of stupidity—one clamorous stupidity, with so many different ways to think it! It was the endless multiplicity of stupidity's manifestations that drove two of the greatest novelistic thinkers of stupidity, Flaubert and Musil, to write unfinished and perhaps unfinishable novels about the problem of *la bêtise* and *die Dummheit*, respectively. These two masterpieces do not support Schopenhauer's claim that laughter emerges in the gap between what we think and what is the case, for that is merely to laugh at our endless errors. Most errors are laughter neutral and that the comedy of stupidity must be tackled not among the dimwitted and the cognitively sluggish, but rather among intellectuals, self-proclaimed philosophers, self-congratulatory experts, and pompous professors.

Planning for the Parallel Campaign, which takes place in 1914, only months before the outbreak of World War I, the central characters in Musil's *Der Mann ohne Eigenschaften*[6] plan to commemorate the coincidence that in 1918 Franz Joseph will have overseen the Austro-Hungarian Empire for seventy years and Emperor Wilhelm II will have ruled Germany for thirty years. Such a coincidence called for a world-historic event, although no one could say what it should be and so endless committees are organized and they are charged with coming up with a totally game-changing idea. Diotima's "soul," perhaps evoking Socrates' fabled teacher in the *Symposium*, feels trapped in both a loveless marriage and a hopelessly materialistic age and it consequently longs for something truly earth shaking. "We must bring to life a truly great idea. We have the opportunity, and we must not fail to use it" (MWQ, 95). When Ulrich asks her what she has in mind, he learns that she "did not have anything specific in mind. How could she? No one who speaks of the greatest and most important thing in the world means anything that really exists" (MWQ, 95). Such incredible, world-shaking ideas "exist in a kind of molten state through which the self enters an infinite expanse and, inversely, the expanse of the universe enters the self, so that it becomes impossible to differentiate between what belongs to the self and what belongs to the infinite" (MWQ, 114). Only by tossing Austria into the dark night when all cows are black can we mark an occasion like the Parallel Campaign in its proper greatness! But how do we do that?

All this leads to committees, endless committees. In soliciting "great ideas" the Parallel Campaign receive countless entries. This allows Diotima to fawn over Arnheim, the wealthy German businessman and author "who possessed the gift of never being superior in any specific, provable respect but, owing to some fluid, perpetually self-renewing equilibrium, of still coming out on top of every situation." Arnheim called this "the Mystery of the Whole": "in this life, in some mysterious fashion, the whole always takes precedence over the parts" and hence everything is somehow saturated in mystery and so that the "profound goodness and love, the dignity and greatness, of a person are almost independent of what he does," but they still magically ennoble him anyway (MWQ, 207).

Among the many characters in the background of these endless conversations and committee meetings is Moosbrugger, who sits in jail awaiting execution after murdering a prostitute, all the time in a psychotic stupor, extravagantly hallucinating. He was, however, "pleased that he had this knack for hallucination that others lacked; it enabled him to see all sorts of things others didn't, such as lovely landscapes and hellish monsters" (MWQ, 258). When Moosbrugger had a handle on his hallucinations, he would engage "in thinking," although he had no idea what that really was. "He called it thinking because he had always been impressed with the word" (MWQ, 258). It described something that happened to him against his will, as if thinking "were planted in him." And what kind of "thoughts" came to him when he was "thinking"? "A Squirrel in these parts is called a tree kitten . . . but just let somebody try to talk about a tree cat with a straight face!" Moosbrugger was, however, convinced that in Hesse a tree kitten was "called a tree fox. Any man who's traveled around knows such things" (MWQ, 259). Deep down, however, Moosbrugger's "experience and conviction were that no thing could be singled out by itself, because things hang together" (MWQ, 259). It usually took "his enormous strength to hold the world together" (MWQ, 259). Moosbrugger, lost in the madness of the dark night when all cows are black, struggled to protect his tenuous hold on things from their dissolution into pure continuity.

Meanwhile, the search for a world-shaking idea continued as everyone attempted to invoke their powers of genius. The committee members are all fairly bright and they can grasp complex ideas, but they are subject to what Musil dubbed *intelligent stupidity*, a stupidity that may deprive the thoughtless life of its good conscience, but which in its own way understands the problem of stupidity to run deeper than the mere abdication of thinking. "For if stupidity, seen from within, did not so much resemble talent as possess the ability to be mistaken for it, and if did not outwardly resemble progress, hope, and improvement, the chances are that no one would want to be stupid, and so there would be no stupidity" (MWQ, 57). Stupidity's power to become a kind of *Doppelgänger* for thinking means that there is "no great idea that stupidity could not put to its own uses; it can move in all directions, and put on all the guises of truth. The truth, by comparison, has only one appearance and only one path, and is always at a disadvantage" (MWQ, 57). Such stupidity, Musil argued, is not merely "a deficiency of the understanding" (OS, 276): "the higher, pretentious form of stupidity . . . is not so much lack of intelligence as failure of intelligence, for the reason that it presumes to accomplishments to which it has no right" (OS, 283). This produces a catastrophic "disproportion between the material and the energy of culture" (OS, 283). That is to say, intelligent stupidity produces energy far beyond the capacity of its host culture to deploy it wisely. Such energy continues to mount until it is expended with simple but cataclysmic violence. Musil further contended that such stupidity "is no mental illness, yet it is most lethal; a dangerous disease of *Geist* that endangers life itself" (OS, 284).

Before Musil, who had a doctorate in philosophy but who had refused an academic position in order to live in penury while he worked on his seemingly unfinishable novel about the problem of stupidity, Flaubert's posthumously published *Bouvard et Pécuchet* (1881)[7] demonstrated that he, too, had been obsessed with this delicate, but inherently catastrophic problem. This novel, also unfinished and perhaps unfinishable, chronicles the relentless intellectual pursuits of the title characters, two former Parisian copyists (i.e., those who, like a parrot, merely repeat, unburdened by the desire to understand what they are repeating). When Bouvard's uncle dies, he inherits a substantial sum of money and the duo soon move out to the country and devote themselves to their newfound intellectual pursuits, which, in a series of disasters, move through what they imagine to be agriculture, landscape gardening, food preservation, anatomy, medicine, biology, geology, archaeology, architecture, history, mnemonics, literature, drama, grammar, aesthetics, politics, love and sex, gymnastic training, occultism and esoteric spirituality, theology, philosophy, religious life, education, music, and urban planning before they return, with an immense sense of relief, to the life of copyists.

Jorge Luis Borges cites Emile Faguet's exasperated complaint in 1899 that "Flaubert makes them read an entire library *so that they will not understand it*."[8] Indeed, Flaubert had worked through fifteen hundred often-difficult tomes to be able to dramatize Bouvard and Pécuchet's *bêtise* with regard to these ideas (DBP, 387). Flaubert had to in some way understand a vast range of material in order to locate and write about with precision the many terrains of his two protagonists' *bêtise*. He attempted to locate surgically stupidity's ill will. Just as Musil's *Man without Qualities* is enormous, yet nowhere near finished at fifteen hundred pages, so obsessed was it with exploring the seemingly endless variety of forms of human stupidity, Borges notes that the late Flaubert was like Swift: "Both hated human stupidity with minutious ferocity; both documented their hatred with trivial phrases and idiotic opinions compiled across the years" (DBP, 388). It is not without irony that in his *Dictionnaire des idées recues*, a collection of habitual exercises of *bêtise*, which Flaubert developed as a supplement to *Bouvard et Pécuchet*, we find the following definition for IMAGINATION: "Always 'lively.' Be on guard against it. When lacking in oneself, attack it in others. To write a novel, all you need is imagination."[9] If all you need is imagination to write a novel, then you cannot understand what the novel, on its own terms, seeks to understand. Bouvard and Pécuchet, personifications of *bêtise*, cannot write a novel or even a philosophical treatise about *bêtise* because they cannot stand in thoughtful relation to it, even though what thinking stands in relationship to in such encounters threatens to stupefy thinking.

And what of Flaubert, who both writes a novel and seeks to understand his characters as they are mired in an uncomprehending *bêtise*? Derrida has alerted us to Flaubert's 1875 letter to Edma Roger des Genettes where he exclaims "Bouvard and Pécuchet fill me up to the point that I have become them! Their *bêtise* is mine and it's killing me!"[10] If one understands either the novelistic or the philosophical account of *bêtise* to be wholly uncontaminated by *bêtise*, somehow wholly evading it, standing above it, purely one's opposite as one reigns down on it, excoriating and capturing it, then paradoxically one cannot understand *bêtise*. Flaubert, Derrida claims, was "scared by the *bêtise* that he has made and put to work, by the *bêtise* he was intelligent and *bête* enough to secrete in order to see, to see it, him too, and no longer tolerate it" (BS, 159). *Bêtise* is not the idiocy of everyone who is supposedly not as bright as myself.

Let us return to Bouvard and Pécuchet as they take their new inheritance and move out to the country and try their hand at farming. Rather than seek the advice of the farmer already there working the land, they decide to figure out for the agricultural sciences for themselves. After many autodidactic disasters, they come upon the following hypothesis: they had not sufficiently amended the soil:

> Egged on by Pécuchet, he had a frenzy for manure. In the compost trench were flung together boughs, blood, entrails, feathers—everything that could be found. He employed Belgian dressing, Swiss fertilizer, lye, pickled herrings, seaweed, rags; he sent for guano, and tried to manufacture it; then, pushing his tenets to the extreme, would not let any urine be wasted. He suppressed the privies. Dead animals were brought into the yard with which he treated the soil. Their carcasses were scattered over the country in fragments. Bouvard smiled in the midst of the stench. A pump, installed in a tumbrel, spurted liquid manure over the crops. To those who put on airs of disgust he said: "But it's gold! It's gold!"
>
> And he regretted not having still more dung-heaps. Happy is the country where are natural grottoes filled with bird-dropping! (BP, 50)

Needless to say, the harvest was not robust: "the rape was meager, the oats poor, the corn sold badly because of its smell" (BP, 50).

Their "studies" continued unabated and when, after many other intellectual debacles, they are confounded by literature, they turn to the study of aesthetics, even reading Schelling, who teaches them that beauty is the "infinite expressing itself in the finite" (BP, 166). The new academic enterprise goes into crisis before it can come into itself, foundering on the duo's inability to distinguish successfully the beautiful from the sublime:

> A character is beautiful when it triumphs, sublime when it struggles.
>
> "I understand," said Bouvard, "beauty is beauty, the sublime is great beauty. How can they be distinguished?"
>
> "By intuition," replied Pécuchet.
>
> "And where does intuition come from?"
>
> "From taste."
>
> "And what's taste?"
>
> It was defined as a special perception, swift judgment, the gift of perceiving certain relationships.
>
> "So taste is taste; and all that says nothing about how to acquire it."

> It is necessary to observe the proper rules, but these rules vary: and however perfect a work may be, it will not be altogether free from reproach. Still, there is an indestructible beauty, whose laws we do not know, for its origin is a mystery. (BP, 167)

When their gardens fail, they conclude that "arboriculture is probably all humbug," just "like agriculture!" (BP, 59). And then there is their spastic political involvements, in which their commitments run the gamut from solidarity with the people of the 1848 revolution to a defense of Napoleon—"Let him gag the mob, stamp it under foot, crush it! That will never be too great a penalty for its hatred of the right; its cowardice, its ineptitude, its blindness!" (BP, 203). Exasperated by all of this floundering in the world of ideas, they no longer pursued their studies "for fear of disillusionment" (BP, 204); "sometimes they opened a book, only to close it again—what was the use?" (BP, 204).

Their experiments and forays into animal magnetism and other forms of occultism eventually led them to a philosophical quandary: ecstasy depends on a material cause, but what is matter and what is spirit (BP, 240)? This led them to the study of philosophy and what is now called the mind-body problem. Trying to understand Spinoza and concluding that the universe "is impenetrable to our consciousness" (BP, 244), they were quickly overwhelmed. "All this was like being in a balloon at night, in glacial coldness, carried on an endless voyage towards a bottomless abyss, and with nothing near but the unseizable, the motionless, the eternal. It was too much. They gave it up" (BP, 244).

They turned to philosophy primers, but these, too, proved vexing until "both avowed their weariness of philosophers. So many systems serve only to confuse. Metaphysics has no use. One can exist without it" (BP, 249). Still, despite pressing quotidian demands, metaphysics returned as "thoughts had come bubbling up" (BP, 250). All this was to no avail. Although "they began reasoning on a solid basis . . . suddenly an idea would vanish, as a fly darts off when one tries to catch it" (BP, 250–251). Pécuchet turned to Hegel, but soon Bouvard was exasperated: "One explains what one knows little about, by means of words that one doesn't understand at all" (BP, 255). This soon drives them to the depths of nihilism: "The certainty that nothing exists (however much to be deplored) is not the less a certainty. Few people are capable of holding such a belief" (BP, 255). Soon Bouvard is defending even rogues. "When a man's born blind, or an idiot, or a murderer, that seems disorder to us, as though order were known to us, as if nature acted with a purpose" (BP, 257). Bouvard and Pécuchet continue to routinely undermine "all foundations" (BP, 258).

Although they are threatened with prison and slandered for being immoral, something much worse transpires. In their heroic nihilism, something strange begins to emerge, something to which Deleuze also alerts us:

> the presentiment of a hideousness proper to the human face, a rise of *bêtise*, an evil deformity, or a thought governed by madness. For from the point of a philosophy of nature, madness rises up at the point at which the individual contemplates himself in this free ground,—and, as a result, *bêtise* in *bêtise*, cruelty in cruelty—to the point it can no longer stand itself. (DR 197/152)

As we discuss shortly, Deleuze admits in a footnote that this kind of articulation of the problem borrows from Schelling's *Freedom* essay (DR 198/321–322), but at this point he also alludes to *Bouvard et Pécuchet* when Flaubert famously tells us that "then a pitiable faculty developed in their spirit, that of perceiving *bêtise* and no longer tolerating it" (BP, 258; DR 197/152).

What comes of this "pitiable faculty"? The *bêtise* that imagines everyone else to be idiotic, grows to spite itself and then succumbs to melancholy, the melancholy that Schelling in the *Freedom* essay spoke of as the attractive force of the reemerging abyssal ground that pulls the stupidity of self-grounded ideas into its vortex: "The human never receives the condition within their power, although they strive to do so in evil. . . . Hence, the veil of melancholy that is spread out over all of nature, the deep, indestructible melancholy of all life" (I/7, 399).[11] Poor Bouvard and Pécuchet sink deep into the burgeoning evil and flaring illness of their aggressive *bêtise*:

> Insignificant things made them sad; advertisements in the newspapers, a smug profile, a foolish remark heard by chance . . . they felt as though the heaviness of all the earth were weighing on them. They no longer went out, or received visits.

> . . . The word was diminishing in importance; they saw it as through a cloud, come down from their brains, over their eyes. (BP, 258–259)

In the *Dictionnaire* we learn under the entry TO THINK: "painful. Things that compel us to think are generally neglected" (DAI, 79). This is the evasion of thinking that hides, initially and for the most part, in the *bêtise* of what passes for thinking, much in the way that Franziska "thought" that she was consoling her son by telling him that his "books are now world shaking." This is the kind of "thinking" that manifests in the *Dictionnaire* entry for MISTAKE: "'It's worse than a crime, it's a mistake' (Talleyrand). 'There is not a mistake left to commit' (Thiers). These two remarks must be uttered with an air of profundity" (DAI, 58). The problem with this entry is not that it is a mistake about mistakes, but rather that it protects thinking ("painful") from the "stupidity" of its ground by making oneself the ground of thinking. This reversal or inversion—the force of illness, evil, madness, and melancholy—does not originate in myself nor is it by simply dwelling in the ground that we shun in our anxiety before the possibility of being stupid. In English, the word "stupid" derives from the Latin *stupidus*, to be "confounded and amazed," itself derived from the literal meaning of *stupidus* as "struck senseless," from *stupere*, to be stunned and confounded, and hence retaining a relationship, which was preserved in English until sometime in the eighteenth century, with the word "stupor," itself the result of being paralyzed by surprise. *Dumm*, whose origins are less clear, suggests inexperience, but seems also to be related to *stumm*, to be mute, as if the experience of stupidity swallowed up all other experience, leaving one with nothing really to say, unable really to use language. It is to be left speechless in the sudden muting of what one has imagined to be one's experience.

Such unexpected muting itself cannot be wholly separated from death, and when Bouvard and Pécuchet joylessly walk the countryside they come upon the corpse of a dog.

> The four legs were dried up. The grinning jaw revealed ivory fangs beneath blue shops; instead of the belly there an earth-colored mass that seemed to quiver, so thickly did it pullulate with vermin. It stirred, beaten by the sun, under the buzzing of flies, in that intolerable stench—a fierce, and as it were,

> devouring odor. . . . Pécuchet said stoically: "One day we shall be like that." (BP, 260)

They were under the sway of the idea of death, and "they tried to imagine death in the form of a very dark night, a bottomless pit, a never ending swoon—anything at all was better than this monotonous, absurd and hopeless existence" (BP, 260). They considered what they had missed out on. Bouvard yearned for wealth, wine, and "beautiful yielding women," while Pécuchet dreamed of mastering philosophy so that he could solve all problems by solving a problem so vast that it contained, and thereby simultaneously resolved, all other possible problems. And then they contemplated suicide, but eventually averted it by discovering religion. Facing the silent stupor of the noose, a song about the King of Angels made them feel as if "a dawn were rising in their souls" (BP, 264). Religious life brings them back from the abyssal stupidity of death, and soon they are back at work "thinking" in their serial progression of intellectual catastrophes about education, music, and urban planning.

How does *bêtise* so easily escape the stupefying silence of the noose and remain so "unshakable" such that, Flaubert reflected, "nothing can attack it without breaking against it" (quoted in BS, 160)? *La bêtise* is as relentless as death itself, much like Sancho Panzo characterized it in Cervantes masterpiece:

> [Death is] more powerful than finicky; nothing disgusts her, she eats everything, and she does everything, and she crams her pack with all kinds and ages and ranks of people. She's not a reaper who takes naps; she reaps constantly and cuts the dry grass along with the green, and she doesn't seem to chew her food but wolfs it down and swallows everything that is put in front of her, because she's as hungry as a dog and is never satisfied. (DQ, 590)

As such, it drives us from its ground to the violent safety of ourselves as the unquestioned authors of fixed positions, self-assured answers, and comfortable conclusions.

Borges admired Flaubert's sense of this in the latter's appreciation of Spencer's critique of science as "a finite sphere that grows in infinite space; each new expansion makes it include a larger zone of the unknown, but the unknown is inexhaustible." Hence, Flaubert concluded that "we still now know almost nothing and we would wish to divine the final word that will be never revealed to us. The frenzy for reaching a conclusion is the most sterile and disastrous of manias" (DBP, 388). Let us be clear: the violence of *la bêtise* lies in a madness, but a very particular one and one that is supremely destructive, namely, the insanity of the self-grounded ego to still the ground of thinking with grand conclusions, lest it stupefy one with the abyssal force of death itself. Was this not Pécuchet's greatest desire, standing before the noose, of utterly mastering philosophy so that he could solve all problems by solving a problem so vast that it contained and thereby simultaneously resolved all other possible problems? Derrida picks up on the same concern. In his critique of Auguste Comte's positivistic *bêtise*, Flaubert warns that "ineptitude consists in wanting to conclude. . . . It is not understanding twilight, it's wanting only noon or midnight. . . . Yes, *bêtise* consists in wanting to conclude" (BS, 161). *Bêtise* is the rule of the result.

The Drive for Conclusions

The mood of Flaubert's declaration that "*bêtise* consists in wanting to conclude" shimmers through Schelling's account of philosophical errors not as the incapacities of intelligence, but as the moments in which thinking inhibits itself, clots, refuses to go on, stops cold. "There are no universally valid propositions, only propositions that are valid for *the* moment of development of which they are an expression."[12] Ideas do not have free-floating veracity lying about, ready to be picked up and contemplated. They must be thought genetically, in their becoming within a domain in which they can have meaning. Deleuze, in this Schellingian vein, argues that "all determinations become bad and cruel when they are grasped only by a thought that invents and contemplates them, flayed, and separated from their living form, adrift upon this barren ground. Everything becomes violence on this passive ground." This for Deleuze is the site of the "Sabbath of *bêtise* and malevolence [*méchanceté*]" (DR 198/152). On the Sabbath, the genetic dynamism of the ground is given a rest in the malevolence of conclusions that allow the ground to be usurped by a thinking that cannot affirm it and, in failing to do so, takes refuge in itself as the ground, thereby inverting the original relationship.

The anxiety that the threat of *bêtise* evokes is the threat of a ground whose life cannot be settled in final conclusions. Hence, in the *Freedom* essay, Schelling famously argued that "the anxiety of life itself drives the human from the center in which they were created" (I/7, 382).[13] Thomas Buchheim alerts us to the fact that this phrase, *die Angst des Lebens*, can be taken in both the *genitivus subiectivus* and the *obiectivus* forms.[14] In the objective sense of the genitive, it would be a shorthand way of expressing *die Angst vor dem Leben*, the mounting anxiety in the face of what one experiences as the mounting stupefaction of life, Bouvard and Pécuchet when "they opened a book, only to close it again—what was the use?" It is as if, Schelling tells us, we were to stand at a great height and suddenly, as if from nowhere, we felt tempted to jump. In the subjective sense, however, one could say that this is the anxiety that belongs to the movement of life—life's own anxiety, its inability to decide between its two contesting centers, *Grund* and *Dasein*, ground (which Schelling radicalizes, following Böhme, as *Ungrund*) and presence. Anxiety is the governing affect that corresponds to the conflict of directions in Being, because it does not know whether to go in or out, to live or die.

This is the anxiety at the heart of stupidity. It is not something in itself (hence the endless variation of possible forms of stupidity and Flaubert's exhaustion and Musil's gargantuan but still unfinishable novel). It is, as Deleuze argues, following Schelling, "neither the ground nor the individual" but rather the relation in which the relation between ground and presence is reversed and hence it is "this relation in which individuation brings the ground [*fond, (Un)grund*] to the surface without being able [*sans pouvoir*] to give it form (this ground rises by means of the I, penetrating deeply into the possibility of thought and constituting the unrecognized in every recognition)" (DR, 197–198/152). The I, anxious before a ground that consumes the very determinations that it births, confronts it as a dark night when all cows are black and flees to its periphery by fleeing into oneself as ground. This, Deleuze tells us, is what is "the most tempting in the most stupefied moments [*les moments de stupeur*] of an obtuse will. For this ground, along with the individual, rises to the surface yet assumes neither form nor figure. It is there, staring at us, but without eyes" (DR, 197/152).

Michel Foucault very much admired Deleuze's courage with regard to this problem, arguing that "within categories, one makes mistakes; outside of them, or beyond or beneath them, one is stupid. Bouvard and Pécuchet are acategoriacal beings."[15] When they cannot complete a thought, they immediately dismiss it as hopeless. "'It's the same either way' stupidity says" (TP, 362). It is catatonic stupefaction in which all thoughts are arbitrary. Drugs, which unless you are a fortune-teller, have nothing whatsoever to do with the problem of truth and falsity—they, too, are acategorical, but unlike stupidity, they are not paralyzed by the dark night when all cows are black—"displace the relative positions of stupidity and thought by eliminating the old necessity of a theater of immobility" (TP, 363). As Deleuze, following Schelling, puts it, Flaubert's "pitiful faculty" can become the "royal faculty" as it renders possible "a violent reconciliation between the individual, the ground, and thought" (DR, 198/152). The royal repetition, when the groundlessness of death is longer the stupefaction of the dark of the black cows, is, as we have seen in the previous chapter, "a belief in the future [*croyance en l'avenir*]" (DR, 122/90). Death becomes the productivity of thinking. "In this manner, the ground has been superseded by a groundlessness, a universal ungrounding which turns upon itself and causes only the yet-to-come to return" (DR, 123/91).

Stupidity flees the ground as if it were the mere opacity of death, the dark night when all cows are black, but this is the violence in which the ground of life is bracketed by the dogmatic image of thought, as mere stupefaction. Hence, I agree wholeheartedly with my colleague Julián Ferreyra that, in contradistinction to the kind of Deleuzian who drunkenly rushes, intoxicated by rhizomes and promises of lines of flight, to an undifferentiated abyss as if they were glibly not as stupid as Bouvard and Pécuchet, "becoming a philosophy of indifference is the greatest risk for the philosophy of difference."[16] How stupid to have spoken of the dark night! *Bêtise* is not the stupefaction of utter indeterminacy, that is, Moosbrugger's insane experience and conviction that no thing could be singled out by itself because at the level of ground all things hang together in pure continuity. As Deleuze argues in the conclusion to *Difference and Repetition*, defending Schelling from Hegel's charge of the night of the black cows, the illusion internal to all representation is that "groundlessness (*le sans fond*) should lack differences, when in fact it swarms with them" (DR, 355/277). As the I fractures, the abandoned relationship to the *Ungrund* reasserts itself and we succumb to melancholy and malevolence or, as Ferreyra notes, "we attain the point of view of genesis" and "ideas spring in the apparent groundlessness of *bêtise*."[17] As Deleuze had already argued in 1962, Nietzsche had made active thought a "critique of *la bêtise* and baseness [*la bassesse*]" and in so doing, proposed a new image of thought, liberated from the malevolence and stupidity of the dogmatic image of thought. "Thinking depends on forces which take hold of thought. Insofar as our thinking is controlled by reactive forces, we must admit that we are not yet thinking" (NP, 123/108).

This was the guiding insight into Schelling's classification of our species into three fundamental types. In the third draft of *Die Weltalter* (1815), he dissects the problem of the *Verstandesmensch* (the "intellectual") as a problem of stupidity:

> One could say that there is a kind of person in which there is no madness whatsoever. These would be the uncreative people incapable of procreation, the ones that call themselves sober spirits. These are the so-called intellectuals [*Verstandesmenschen*] whose works and deeds are nothing but cold intellectual works and intellectual deeds. Some people in philosophy have misunderstood this expression in utterly strange ways. For because they heard it said of intellectuals that they are, so to speak, low and inferior, and because they themselves did not want to be like this, they good-naturedly opposed reason [*Vernunft*] to intellect instead of opposing reason to madness. But where there is no madness, there is also certainly no proper, active, living intellect (and consequently there is just the dead intellect, dead intellectuals). For in what does the intellect prove itself than in the coping with and governance and regulation of madness? Hence the utter lack of madness leads to another extreme, to imbecility (idiocy), which is an absolute lack of all madness. But there are two other kinds of persons in which there really is madness. There is one kind of person that governs madness and precisely in this overwhelming shows the highest force of the intellect. The other kind of person is governed by madness and is someone who really is mad. (I/8, 338–339)

Stupidity is the lack of madness at the heart of knowing, a refusal of the concept as chaoid. It knows things from the periphery, mastering their forms while sublimating their elemental monstrosity. Thinking finds itself before the crossroads of two great violences: the stupidity of dogmatism (the malevolence, incipient or manifest, of the self-assured dogmatist) and the dark night of ungoverned madness. Philosophy negotiates madness as its vocation, with all due modesty, in its attempt to keep the upper hand on stupidity and stop being the only beast that can be beastly.

By the phrase "keep the upper hand," I would like to evoke Heidegger's argument that one cannot overcome (*überwinden*) metaphysics, but rather that one must do one's best to "keep the upper hand [*verwinden*]" on it. I do not disagree with the danger of metaphysics, but I hope that I have successfully suggested that the problem of metaphysics cannot be extricated from the more fundamental problem of the violence of stupidity. This requires the kind of double gesture that has been the transcendental concern of this chapter and a key to Schelling's image of thought, a gesture in which *die Verwindung* negotiates the twin violences of metaphysical dogmatism and melancholic paralysis. It cannot fully evade the two, but the dignity of philosophy lies in the ongoing practice of not being utterly consumed by them.

There are a host of issues regarding the metaphorical animality at the heart of the problem of *la bêtise*. *L'animal*, Deleuze tells us, is "protected" from *la bêtise* "by specific forms" (DR, 196/150) and that "animals are in a sense forewarned against this ground, protected by their explicit forms" (DR, 197/152). The human is the only *bête* with a faculty (in the Kantian sense) of *la bêtise*, the only *bête* that can be *bête*, the only beast capable of being beastly. This is perhaps also a reference to Schelling's claim in the *Freedom* essay, following Franz von Baader, that the human can "only stand above or below animals" (I/7, 373). Perhaps the language of above and below does not take us far enough away from the history of the human subjugation of animals, even though both Schelling and von Baader here link such a history with the human being finding itself beneath animals, that is, having a beastly relationship to beasts. Only human beings can be evil and only human beings can be stupid. That being said, the claim is not that animals are lost in some kind of permanent stupor, dazzled in the nonphilosophical default mode of species specific natural stupidity. The burden is rather on our species to stop having such a *stupid—and I daresay evil*—relationship to our nonhuman animal sisters and brothers.

As Whitehead lamented in *Science and the Modern World*, the stupidity of our age is characterized not by thoughtlessness, but by the "restraint of serious thought within a groove. The remainder of life is treated superficially, with the imperfect categories of thought derived from one profession."[18] As a result we are directionless, imbalanced, with the whole "lost in one of its aspects" (SMW, 246). As the violence that Musil's endless committee meetings could not foresee, let alone forestall, teaches us, unless we claim our right to the problems of thinking, this explosive force continues to acquire catastrophic energy. Religion cannot intervene if it, too, is claimed by stupidity (as it has been for much of its history). Its capacity for emancipatory awakening must not fail to struggle to keep its upper hand on its proclivity (in the Schellingian and Kantian sense of radical evil)[19] for devastating—and devastatingly dogmatic—violence.

The Drunken Sobriety of Philosophical Religion versus the Great Ocean of Stupidity

As we saw in the first chapter, the "necessary evil" of reflection is the unavoidable risk that in cracking the spell of presence and exposing, by way of wonder and doubt, the realm of nature as the realm of images, we nonetheless remain at the periphery of nature (*Ideas*, I/2, 12–13). At the periphery, we conclude that the subject is thinking itself and that nature, emerging in the interrogative attitude as a distant and unknown object, looms detached before us, alien to us, and recoverable only through "objective" investigation and interrogation. This stance (the thinking subject researching separate and distinct objects) is the impossibility of nature itself emerging as a question. "The entirety of modern European philosophy has, since its inception (in Descartes) the shared deficiency that nature is not present to it and that it lacks nature's living ground" (I/7, 356–357).

Why then do we languish on the periphery? Why is philosophy in the sense of Schelling's reactivation of its creative autogenesis so rare? Why does the utopian dimension of Schelling's fundamental philosophical gesture seem so fragile before the prevailing forces of stupidity? If only the problem of stupidity were the problem of cognitive deficiencies! For Schelling, the transcendental problem of stupidity cannot be altogether separated from the problem of evil, that is, when confronted with our fundamental nothingness (our lives do not mean anything beyond themselves), we are driven in *die Angst des Lebens* (I/7, 383) from the center of life to the periphery of life. No being has its being in itself, although humans in evil strive to become their own center. The gravitational center, in itself death and formlessness, is also animating: "Now as the copula [*das Band*] as gravity negates the bonded as existing-for-itself, it equally negates that rest, the nothingness of which we have examined in space, while it sets rest into motion. Motion in rest is therefore in the individual the expression of the *Band*, insofar as it is gravity, that is, identity in totality" (I/2, 366). Gravity is already a prefiguring of the Higgs boson as the *underlying vacuum* at the heart of the visible face of matter. As the animating-besouling of matter, antimatter at the heart of matter is what the Mahāyāna tradition called emptiness or *śūnyatā*, a being's inability have its own being; a being is not itself by itself, having no *svabhāva* or self-being.

Rather than having one's own being while simultaneously fulfilling it in an even greater being (God), Schelling's God is in itself a vacuum whose gravitational pull denies beings any integrity (self-being). As we have seen, this is not bad news, but rather the portal to the creativity of the unfolding "cosmic poem." The force of life drives human life from the center of its own life. "Most people turn away from what is concealed within themselves just as they turn away from the depths of the great life and shy away from the glance into the abysses of that past which are still in one just as much as the present" (I/8, 207–208). From the periphery become center, the war of human subject against subject ensues:

> Whenever the light of that revelation faded and humans knew things not from the All, but from other things, not from their unity but from their separation, and in the same fashion wanted to conceive themselves in isolation [*Vereinzelung*] and segregation from the All, you see science desolated amid broad spaces. With great effort, a small amount of progress is made, grain of sand by grain of sand, in constructing the universe. You see at the same time the beauty of life disappear, and the diffusion of a wild war of opinions about the primary and most important

> things, as everything falls apart into isolated details [*alles in Einzelheit zerfallen*]. (I/7, 140)

Schelling had written the *Freedom* essay to oppose the "sectarian spirit that too often rules [*zu oft beherrschender Sektengeist*]" German thought. Philosophy is called to the *Ungrund* that is the source of our difference as precious and cherishable. Rather than partition, it demands a *kunstreiche Ausbilding*, a formation, education, even a cultivated imagination that is artistic, ingenious, literally rich in art, and not, as usually prevails, *die künstliche Schraubengänge der Polemik*, the artificial screw threads of polemic. One could not even say that the *künstlich* is poor in art. It is opposed to art, a mere shadow of art that allows the enemies of art to pose as art. Hermann Broch, as we see in the next chapter, called this *kitsch* and considered it to be a form of radical evil. In the *Philosophy of Art* lectures, Schelling argued against what he dubbed the prevailing "literary Peasant Wars [*Bauernkrieg*]," alluding to the 1524 to 1525 failed peasant revolt for freedom against the Feudal establishment; freedom is not Hobbes' war of all against all, difference as "the diffusion of a wild war of opinions about the primary and most important things" (the darkly violent side of the postmodern condition).

Despite the possible limitations of his political stance, Schelling remained suspicious of the power of the State, arguing that peace cannot be imposed from without. As late as 1850 Schelling still insisted, as he had in the *System* fragment,[20] that no perfect state was possible and that "its task is to provide the individual with the greatest freedom (*autarky*)—freedom, namely, through the state beyond the state [*über den Staat hinaus*], to be, so to speak, beyond the state" (II/1, 551). Hence *Ausbildng* is utterly antithetical to what Hitler called *Gleichschaltung*, the shoving and forcing into sameness and bringing into line (if difference is destroyed, there will be no resistance). "This knowledge is not a light that simply illumines from without, but rather one that excites from within and moves the whole mass of human cultivation [*Bildung*]" (I/7, 141).

The eternal or peripheral view is what Dostoevsky's underground man, despite being invisible to virtually all human beings with whom he came into contact, memorably called the *right to be stupid*. Arguing that no one would surrender their willful freedom to a preplanned, reasonable, good life, we would irrationally revolt against "some well-meaning publications . . . in which everything will be so precisely calculated and designated that there will no longer by any actions or adventures in the world."[21] For the underground man, more important than human contact is the right to declare that 2 + 2 = 5. This is not, as has long been argued, some dramatic existential declaration in which we affirm the precious gift of our freedom. This is rather an act of the bellicose will whose violence stems from the anxious assumption that we are the subject and freedom is the predicate. Freedom of the right to be stupid is not the devotion to the way of things, but rather the right to assert one's will against nature, to assume that we are free and that nature is a mere coefficient of resistance. (Fichte never fully solved this problem and it is at the heart of their rupture in 1802.[22])

Hence, after a humiliating evening in which he is utterly ignored, the underground man does not seek the companionship of which he is so destitute, but rather, sensing that the prostitute who has come to visit him is even further down the chain of recognition than he, he lashes against her. "We're stillborn, and have long ceased to be born of living fathers, and we like this more and more. . . . Soon we'll contrive to born somehow from an idea" (NU, 130). Is this perhaps the sublimated wish of all *Verstandesmenschen*? Not only to live in the world of ideas, but to be born from them?

What could be further away from the Schelling who declares a religious love of things just as they are?

> Religion, however, is the contemplation [*Betrachtung*] of the particular in its being bound up with the All. It ordains the natural scientist [*Naturforscher*] as a priest of nature through the devotion [*Andacht*] with which she cares for the individual details. Religion assigns to the scientist the limits, posited by God, of the drive for the Universal, and thereby mediates science and art through a sacred copula [*heiliges Band*]. (I/7, 141)

Schelling's freedom was the freedom of nature that is testified by art (itself a form of controlled errancy). We can create not because we are willful but because we *surrender* to the creative process. The underground man, to the contrary, demands "*to have the right* to wish for himself even what is stupidest of all and not be bound by an obligation to wish for himself only what is intelligent" (NU, 28). This is the violence at the heart of the dogmatic image of thought, its transcendental imperative to administer the whole of the cosmos from its periphery, to tame the imagination by equating it with one of its presentations.

It is in this respect that we can appreciate the singularity of Schelling's embrace of philosophical religion. Using a genealogical critique, he wrests its hidden treasure from the ruins on which subsequent stupidities were constructed. For Schelling religion has nothing to do with a stupid, utterly arbitrary, and philosophically embarrassing commitment to believe in some kind of transcendent thing or another—"immanence and transcendence are completely and equally empty words, since even they precisely suspend this contradiction, and everything flows together into one, God-filled world" (I/2, 377).[23] Thinkers like Eschenmayer, following a long tradition, sought God in the heights, but Schelling sought God in the depths (I/8, 163). There is no other life than this life, but it is all that we could need or want. Nonetheless, the stupidities of religion besiege us from all directions. Beyond the stupor of a religious inability to pose questions and pursue them creatively, or its idolatrous attachments to transhistorical ideas, or its refusal of other religions for the simple reason that they are not one's own, or its ongoing humiliation of philosophy as a mere means to drum up evidence and rationales for preestablished conclusions, or its ongoing, anxiously aggressive siege against science, Schelling insists that the philosophical question of religion—the secret of the scientist's love of nature as much as the artist's surrender to creativity—should not live and die on the periphery where stupidity reigns and we fight for our deaths as if it were for our lives.

Richard Dawkins is right to strike back with *The God Delusion* and Stephen Jay Gould commands our admiration for taking on the creationists, but it is equally right to say that it is not the vocation of either science or philosophical religion to expend its precious resources on a fight that is not at its heart philosophical, rational, or, in any defensible sense, thoughtful. Schelling, for whom God is a question throughout his entire path of thought, is in his own way a militant atheist if by that one means that there is such a *thing* or *existing being* as God. There is no transcendent object that endows from afar His creation with intrinsic meaning, and when this meaning comes into conflict with the natural sciences, automatically trumps them. In this respect, Schelling would have to accord to a point with Dawkins' militant atheist stance about meaning, namely, that "there is something infantile in the presumption that somebody else (parents in the case of children, God in the case of adults) has a responsibility to give your life meaning and point."[24] Nonetheless, Dawkins does not fully extricate himself from the Platonism (the sensible realm as the mere appearance in time of the eternal intelligible realm). There is no meaning to anything remote from nature.

Although the straw man God that Dawkins slays is widely and often violently prevalent, it is nonetheless a straw man God (an intelligible being that somehow exists remote to the apparent world of creation) and as such exemplifies the painful stupidity of religion. Dawkins, in calling for us to take responsibility for the meaning of our own lives and take the "truly adult view . . . that our life is as meaningful, as full and as wonderful as we choose to make it,"[25] inadvertently falls into the trap of the reverse Platonism that one also finds in Sartre. The lack of a priori meaning charges us with creating our own a posteriori meaning. We are what we make of ourselves. To call the obverse of this view "infantile," however, is not fair to children, who are perfectly happy with the absurdity of their lives (children do not play because it makes their lives more meaningful). They must be taught that happiness is allegedly dependent on meaning, and in order to flourish we must either find that meaning (Platonism) or make something out of ourselves (reverse Platonism). It also condemns adults to the periphery, looking at themselves as objects to endow with meaning and at the earth as an object to analyze.

To put it provocatively: Schelling is, paradoxically, a *religious atheist*. The idea of God is not the idea of any sort of a thing; God names the unthinged, *das Unbedingte*. The challenge, therefore, is neither to find meaning remote to the things themselves nor to administer meaning from the periphery of our lives and the earth. Philosophy and religion name the rituals by which, as Dōgen (1200–1253) said in his superb fascicle *Genjōkōan* (現成公案) (1233), "to study the Buddha way is to study the self; to study the self is to forget the self; to forget the self is to be confirmed by all things."

Schelling presents his own version of this three-part movement in his Berlin lecture on the *Philosophical Introduction to the Philosophy of Mythology*.[26] Taking up a distinction found in the Catholic archbishop François Fénelon (1651–1715), himself accused of Quietism, the first part of the movement consists of the self seeking to forget itself and to abnegate itself—in essence to remove oneself from the periphery. "The ego seeks to consummate the act of self-forgetting or abnegation of itself. This presents itself in mystical piety . . . which consists in the person seeking as much as possible to void [*vernichtigen*]—but not to annihilate [*vernichten*]—themselves and all the merely accidental being pertaining to themselves" (II/1, 557).[27] The Swabian mystics that had held the young Schelling in thrall did not seek to annihilate themselves, but rather to liberate themselves from the periphery. *Vernichtigen* is to make oneself nothing, void, and invalid, "to count one's self as nought," but in the sense here of emptying oneself of oneself, undergoing the κένωσις at the heart of nature.

To stop here, however, is finally only to have yielded to the ultimate range of all negative philosophy: only to have emptied oneself of oneself, to have abandoned oneself as the subject and thereby to have abandoned the alienation of reflection that is the view from the periphery. Negative philosophy only achieves this intuition of the infinite within the finite, and concludes with the ungrounded ground from which thinking arises. In the Munich lectures on the history of modern philosophy (1827), Schelling warns against the lazy or even pernicious recourse to mysticism as what we would now call *mystification*. In common parlance, the mystical is simply what I do not understand. The "mystical" becomes the magic shield that keeps thinking from becoming aware of its own stupidity.

Mysticism is therefore "a hatred of clear insight" (HMP, 192/185). The term itself, το μυστικόν, however, speaks to the accomplishment of negative philosophy insofar as it simply marks the hidden and concealed; it names the monstrosity at the heart of light, its invisible, disintegrating, gravitational life. In this strict sense we can say that the ground of nature (*natura naturans*) is secret and as such mystical (HMP, 190–192/183–185). This is a move to free science for its own practices of adoration, not to mystically obscure the fruits of scientific research. Philosophy does not drive everything back into the dark night of its birth as an end in itself. It is the opening of a new, nondogmatic image of thought, in which we think *from* prime and not simply *toward* prime.

The second movement, as we see in greater detail in the last two chapters, was already in play in the *Philosophy of Art* lectures (1802–1804). It is to become *like God*—ὁμοίωσις θεοῦ—a phrase that Schelling likely borrows from James 3:9, where humans are spoken of as τοὺς καθ' ὁμοίωσιν θεοῦ γεγονότας, those made in the image of God, the ones made like God. Again as we see in the final two chapters, this is *expressly not to speak of image as a picture that copies the form of an original.* There are no representations of God and we shall make no idols or graven images (famously in Exodus 20:4, but also in Leviticus 26:1, Deuteronomy 4:16–19, and Isaiah 42:8). The Hebrew *pesel* (פֶּ֫סֶל) is a carved or graven copy, and this resonance can also be detected within the roots of the word "image," *imago*, originally naming a copy, statue, or picture (as does the German *Bild*). To copy or represent God as if there were a form to reproduce, or even to imagine that art is the transfer of an original form to its represented form (to copy the look of something in an artistic work, to reduce the imagination of something to the representation of something), is to foreswear the problem of the image for the love of idolatry.[28]

For Schelling there is nothing but the life of images, but to detach images from their living ground is to turn them into dead images, idols. Moreover, their life is in itself unimaginable, a primordial and elementally monstrous image (what Schelling calls an *Urbild* or primordial image) that cannot in itself be seen or heard. The image of thought is an imageless image (nature as the groundless expressivity of all beings, the great clamor being), and the dogmatic image of thought (thinking's representation of itself to itself) is idolatry. The image of the earth of thinking or the Great Ocean of thinking is elemental, but is not in itself any sort of being and is as not having any being (it is *das Sein, als solches*, μὴ ὄν, but not *das Seiende*). *God is the life of the imagination as such*, of the elemental and monstrous ungrounding at the heart of the emergence of all images.[29] The natural and divine imagination, the autogenesis of images from the sovereign depths of nature, becomes human (like to like, ὁμοίωσις) in artistic creation, in which humans reveal themselves also to be *natura naturans*. Humans may not avail themselves of any transcendent meaning to artistic creation as such—it is the event of meaning as such, the movement of the "selfless production" of works (II/1, 557).

Finally, and this is the move of philosophical religion as such, knowing as such becomes contemplative in the sense of Plato's διάνοια that discursively discerns beings as they "exist for their own sake" (II/1, 557), not as proxies for some remote meaning. *Things just as they are*, to paraphrase what the Buddhist tradition calls suchness (*tathatā*), is the sensibility of philosophical religion that affirms and knows all images simultaneously *in themselves and as of the primordial source* (what

Schelling often calls *die Urquelle*). This is an early formulation of the attitude that Heidegger will later link to the relationship between thinking and thanking.

But is this not all mad? Does not Schelling injure the carefully won sobriety of philosophy and introduce an element of madness into the temperate domain of rationality? Whoever thought thinking was at its heart merely sane? That is the provenance of the *Verstandesmensch* or the gravity of Hegel's *Ernst des Denkens*. In a stunning formulation that strikingly anticipates Nietzsche's *Birth of Tragedy*, Nietzsche *avant la lettre* while simultaneously unleashing the thought of God, Schelling at the end of his life again takes this up:

> The mystery of true art is to be *simultaneously* mad and level-headed [*wahnsinnig und besonnen*], not in distinctive moments, but rather *uno eodemque actu* [altogether in a single act]. This is what distinguishes the Apollonian inspiration from the Dionysian. The highest task of art, which gleams before us like a miracle, is to present in the most conditioned and finite form an infinite content, which, so to speak, contests all form. God is in no way the mere antithesis of finitude; God is in no way that which can only be pleased in the infinite, but rather God appears precisely as the most artificial of beings [*Wesen*], seeking finitude and not resting until He has brought everything into the most finite form. (U, 422)

The mystery of Christianity, far from its own historical self-understanding, is the Apollonian radiance of the universe, the cosmos as an unfolding poem.[30] Stupidity is revealed as the urge to dominate the earth, while the religious gesture, with its devotion to the monstrosity of the earth, finds its strength precisely in its capacity to become utterly vulnerable. In this sense, Schelling speaks of unreasonableness at the heart of reason:

> The charge against unreasonableness [*Unvernünftigkeit*] only concerns what wants to give itself as itself as reasonable, but what it does not concern itself with is in itself against reason, which can even be explained from a certain standpoint with respect to the ways of human judgment as a stupidity [*Torheit*]. The most audacious of the apostles, Paul, . . . speaks of a divine

> stupidity, of the weakness of God, which is, however, as he says, stronger than all human strength. It is not given to everybody to grasp the deep irony in all divine ways of acting. (U, 420–421)

Schelling here alludes to Paul, who elsewhere calls himself a moron for God ["μωροὶ διὰ Χριστόν" (1 Corinthians 4:10)]. Earlier in the same letter we find the phrase: because the stupidity of God is wiser than humans and the weakness of God is stronger than humans (ὅτι τὸ μωρὸν τοῦ θεοῦ σοφώτερον τῶν ἀνθρώπων ἐστιν καὶ τὸ ἀσθενὲς τοῦ θεοῦ ἰσχυρότερον τῶν ἀνθρώπων) (I Cor1: 25). The stupidity (*Torheit*, μωρός), what itself ironically appears to the *Verstandesmensch* as moronic, is precisely what demonstrates the stupidity of the stupid: it is not me that is stupid, but rather that which I shun that is stupid.

Stanislas Breton's study of Saint Paul, *The Word and the Cross*, speaks of the *shattering power* of the "*logos* of the cross." Breton, too, takes up this letter, especially the passage at 1 Corinthians 1:18: "The *logos* of the cross is folly [μωρία] for those who are perishing, but to the saved, it is the power of God."[31] The language of the cross, the meontic force of death rendered holy, the otherwise than being that comes to contest being, is, for Breton's reading of Paul, "the death of evidence" that is "the true beginning . . . God has chosen that which is not (μὴ ὄντα) to suspend (or put in parentheses) that which is (τά ὄντα)."[32] The force of the μὴ ὄν, the otherwise than being, the *logos* of the cross as it supplants the domain of the *Verstandesmensch* and its prevailing orders of evidence, as well as the invidious consenses according to which we wage our wars of recognition, presents itself to those who do know this force as moronic, as the stupidity of μωρία. This is the power of the radical or divine stupidity that seeks to redeem the immense hold of human stupidity. To end with a beautiful analect of Confucius (5.21):

> As for Ning Wuzi, when the Dao (道) prevailed in the land, he was wise; when it was without the Dao, he was stupid. Others might attain his level of wisdom, but none could match his stupidity.[33]

Part III

Nature of Art and Art of Nature

5

Plasticity

Those who believe classicism is possible are the same who feel that art is the flower of society rather than its root.

—Barnett Newman[1]

As Claude Lévi-Strauss remarked years ago, the arts are the wilderness areas of the imaginations surviving, like national parks, in the midst of civilized minds. The abandon and delight of lovemaking is, as often sung, part of the delightful wild in us. Both sex and art! But we knew that all along. What we didn't perhaps see so clearly was that self-realization, even enlightenment, is another aspect of our wildness—a bonding of the wild in ourselves to the (wild) process of the universe.

—Gary Snyder[2]

Schelling's remarkable public lecture on the relationship between the plastic arts and nature, first published in 1807 as *Über das Verhältnis der bildenden Künste zu der Natur*, was delivered in Munich in the fall in celebration of King Maximilian I of Bavaria's name day (the same occasion out of which *On the Deities of Samothrace* emerged eight years later). The importance that Schelling attached to the speech is suggested by the fact that he included a version of it with six additional comments in the form of endnotes in the first and only volume of his 1809 *Philosophische Schriften*, placing it right before the first appearance of the *Freedom* essay. This was no mere occasional speech in observance of Maximilian's feast day, but a surprising kind of festival, an explosive kind of feast. Schelling, in his call for the "revival [*Aufleben*]," literally, a coming back to life, "of a thoroughly indigenous art [*einer durchaus eigentümlichen Kunst*]" (I/7, 328), and "rejuvenated life [*verjüngtes Leben*]" (I/7, 328) and of an "art that grows out of fresh seeds and from the root" (I/7, 326), which, "like everything else living, originates in the first beginnings" (I/7, 324) and returns to that which in itself is "without image" (*das Ungebildete*) (I/7, 324). That is to say, he calls for an art that returns to life by returning to the source of art's life. He marked this festival as a *kind of carnival*,[3] that is, as a *saturated progression of masks*.[4] It is the aim of this chapter to develop and defend this seemingly eccentric characterization of the address and in so doing begin to thematize the problem of the imagination.

I begin by briefly reflecting on my decision, perhaps provocatively, to speak of this progression of masks as saturated. Why saturated? The latter term derives from Latin roots indicating a drenching, a filling up and satiating. *Die Sättigung*, with its root *satt*, to be full, clearly speaks to this satiation and in reference to chemistry, Schelling reflects in the *Ideas for a Philosophy of Nature* (1797, rev. 1803), for example, on the complete permeation of the alchemical *menstruum*, prime matter, and a superadded body. Without consummate and reciprocal saturation, either the *menstruum* attempts to dissolve the body or a dissolved body attracts a superadded body. Consummate satiation, the complete interpenetration of energy and form, allows for a perfect mixture, such that nothing more could be added.[5] This (al)chemical saturation in which the formless and form—the imageless (*das Ungebildete*) and the image (*Bild*)—interpenetrate in nature can also, Schelling tells us in the Munich speech, be detected in the germinating upsurge of the work of art, as if it were an unknown and unexpected plant: "the artwork rises up out of the depths of nature, growing upwards with definiteness and delimitation, unfolding inner infinity and saturation [*innere Unendlichkeit und Fülle*], finally transforms itself into charm [*Anmut*] and in the end reaches soul" (I/7, 321). The speech is rife with images of living soil, animating ground, and productive earth, all of which oppose mere surface and land (*Boden*). The seed of art takes its life from the depths of this earth, watered by inspiration. Or one could say that in art one finds the saturation of gravity and light, that is, of the dark, attractive depths of ground or "mysterious night" (I/6, 257) of gravity in its coupling with the expanding clarity of form as light. And although Schelling has "represented" or "imagined" (two possible senses of *vorgestellt*) this movement in its constituent and therefore "separated [*getrennt*]" parts, he insists that in "the act of creation" it is "a single deed" (I/7, 321), a unified progression. It happens of itself, beyond the activity of creating or the passivity of being created, in something like the middle voice of artistic productivity (without anything that acts directly or indirectly as a subject).

In art, the soul, the animating *menstruum* of nature, the eternal beginning, or what Schelling in the Munich speech simply calls *das Wesen*, is saturated with form and form is saturated with the living energy of the soul. Although opposing form, there is no soul separate from form because although form delimits energy, it does so in order to give it life and expression. When one conceives form solely in abstract terms, that is, removed from the sensuous, it appears as if it merely constricts and delimits *das Wesen* because it is inimical (*feindselig*) to it, but form has no independent standing. If form is "only with and through *das Wesen*," how could *das Wesen* feel restricted by what it creates (I/7, 303)? "The determinateness [*Bestimmtheit*] of form is never in nature a negation, but rather always an affirmation" (I/7, 303). Form is not restricting and shaping an otherwise passive ὑποκείμενον; it is the creative movement of the earth, nature naturing (*natura naturans*). Creation is the life of nature expressing itself as life.

In his book on Francis Bacon, Deleuze makes this point in relationship to painting when he explicates Bacon's critique of the action paintings of Jackson Pollack: "The diagram must not eat away at the entire painting; it must remain limited in space and time. It must remain operative and controlled. The violent methods must not be given free reign, and the necessary catastrophe must not submerge the whole."[6] Regardless of how one evaluates Bacon's assessment of Pollack—Pollack understood himself to be a force of nature—the broader point still holds: without saturation there is only kitsch (empty forms posing as art) or catastrophe, the dark night of the *menstruum*. Composition demands saturation and in this sense it cannot be separated from *natura naturans*, the productivity and creativity of nature.

But Schelling asks if we any longer understand what it might mean for us to recognize the classical principle that relates the plastic arts to nature since the Greeks, namely that "art is the imitator of nature [*die Nachahmerin der Natur*]" (I/7, 293). That art should imitate nature is a truism whose roots, despite their nineteenth-century fruits, stretch back to the various receptions of Aristotle, including his *Poetics*, where drama imitates and even completes φύσις, and the *Physics*, especially book beta, where Aristotle explicitly claims that τέχνη imitates φύσις (194a21). Aristotle, however, did not mean that the forms that τέχνη brings forth are mere copies of the forms that φύσις brings forth. At stake in φύσις is the problem of bringing forth, of production as such, and hence τέχνη itself also relates to the problem of production, not to the procedure of representing in artistic forms the same forms that first manifested in φύσις. Art is not the skill of mimicking or producing secondhand copies of the forms of nature. It imitates the productivity of φύσις by becoming productive itself.

This was not, however, the way in which the ensuing tradition understood the problem. Marcus Aurelius, for example, in his *Meditations* evokes this perhaps already terminally ill reading of Aristotle when he reflects that "No nature is inferior to art because the arts [merely] imitate the things of nature [Οὐκ ἔστι χείρων οὐδεμία φύσις τέχνης' καὶ γὰρ αἱ τέχναι τὰς φύσεις μιμοῦνται]" (Meditations, xi. 10). From this perspective, plastic images—perceptual forms or shapes if you will—imitate the forms of nature. Natural forms are represented naturally, and whether art copies them and in so doing either falls short of them or improves them, it fundamentally represents them. Representational art produces formal images, *Bilder*, and the source for these formal images is the *Bilder* of nature itself. Art and τέχνη merely copy the look (εἶδος) of nature but do not partake in its self-presentation.

In taking on this deathly perspective, Schelling evokes the great poet-philosopher Johann Georg Hamann (1730–1788), whose *oeuvre* Schelling in 1827 confirmed as the constant "touchstone" of one's own understanding (HMP, 171/168) and whom Schelling, in the 1809 commentarial footnotes on the speech, calls an *urkräftiger Geist*, a spirit of primordial force (I/7, 294) before then exhorting his colleague Jacobi either to edit the "long hoped for" edition of Hamann's works himself or to commission someone else to do it. After Schelling's arrival in Munich, Jacobi had provided him with access to some of Hamann's more difficult to find works as well as to some of his letters. Schelling's relationship to Jacobi would soon degenerate into acrimony, and the dispute, initiated by Jacobi, but which inspired Schelling's uncompromising response, included contrasting accounts of the import of Hamann's writings.[7] It would be almost two decades before the complete edition of Hamann's collected works finally appeared (edited and published over the course of six years in seven volumes and completed in 1827 by Jacobi and Schelling's colleague at the Bavarian Academy, Friedrich Roth). It is perhaps fitting that, given Schelling's lament about the fate of the arts as itself inseparable from the collapse of our relationship to the natural world into the rigor mortis of positivism, that Hamann appeared so irrelevant to the prevailing intellectual climate. Nonetheless, Schelling, out of deference to the regal occasion, offered a "toned down [*gemildert*]" version of Hamann's scathing comment, referring to him discreetly and indirectly as *der tiefsinnige Mann* (the profound man): "Your mendacious philosophy has already done away with nature and so why do you demand that we should imitate it? So that you may be able thereby to revive your enjoyment by exercising the same violent deed against the students of nature?" (I/7, 294).

In the 1809 commentarial footnotes, an emboldened Schelling included Hamann's more incendiary original formulation: "Your murderously mendacious [*mordlügnerisch*] philosophy has done away with nature and so why do you demand that we should imitate it? So that you may be able thereby to revive your enjoyment by murdering the students of nature?" (I/7, 293). Hamann's choice of words merits careful attention. First of all, you murder artists by murdering the source of art, namely nature. Artists are students of nature, but if you kill nature yet still demand that art imitate it, artists produce dead works and in so doing, art and its artists are destroyed. To be clear: the death of art is a consequence of the death of nature and all of this death amounts to a murder spree.

Hamann's image of murdered nature appears again in Schelling's 1811 fragment from the handwritten remains called *Über das Wesen deutscher Wissenschaft* (*On the* Wesen *of German Science*).[8] Schelling claims that Hamann, that "profound spirit," "felt more deeply than anyone the deathblow [*Totschlag*] to nature through the use of abstractions as well as the utter vanity of his age in its elevation above and domination over nature and in its moral enmity toward it" (I/8, 8). The murder weapons were abstractions—deracinated ideas—, the dubious assertion of humanity's self-importance, and an enmity toward nature that stemmed from regarding it at best as value-neutral, but in no way valuable in itself. The crime was not altogether without its "cloud of witnesses" and Schelling counted Böhme and Hamann as chief among them (I/8, 8).

As we have seen throughout this book, Schelling in the *Freedom* essay characterizes the positivistic representation of nature, or more precisely, its view of nature as representable, as nature-cide, the fatal flaw that epitomizes modernity: "nature is not present to it" for modernity "lacks a living ground [*die Natur für sich nicht vorhanden ist, und daß es ihr am lebendigen Grunde fehlt*]" (I/7, 361). Nature therefore becomes a dead abstraction and its forces become mere repetitions of the same. Natural laws are its inviolable operators, and nature is denied its natality, the power of its sovereign and formless life to produce and bring forth—birth—new forms of life, and nature hence becomes incapable of living, free progressivity, so that it merely repeats what it has always already been, "swiveling," Schelling says in the *Freedom* essay, "in the indifferent circle of sameness, which would not be progressive, but rather insensible and non-vital" (I/7, 345). In this "indifferent circle of sameness," nature is reduced and constricted to a system of laws, that is, to continue to develop Schelling's metaphor, a system of dead representations that, like zombies, become implacable forces of the dead. A year before the Munich speech, in the 1806 *Darlegung des wahren Verhältnis der Naturphilosophie zu der verbesserten Fichteschen Lehre*, written after Schelling and Fichte had decisively broken around the question of nature, Schelling claims that "The moralist desires to see nature not as living, but as dead, so that he can tread upon it with his feet" (I/7, 17). Dead nature is the mere surface beneath our feet, the land that we imagine as being there for us, simply at our disposal.

Language emphasizing the life of the living ground preponderates in Schelling's writings during the middle period. He was in part responding to many of his contemporaries who regarded nature, as he puts it in the Munich speech, as a "dead aggregate of an indeterminable quantity of objects" or as abstract space filled with objects like a receptacle, or as raw materials for extraction and consumption—mere "ground from which one draws nourishment and sustenance" (I/7, 293). This contrasts with the soul of forms, "the living center" (I/7, 296), the "saturation (*Vollkommenheit*) of each thing" not through its "empty, abstracted form" but rather through the "creative life in it, its power to be present [*Kraft dazusein*]," which is lost on those who cannot see the genesis and production of form, but rather just see "nature overall as something dead" (I/7, 294). Life is chemical saturation in the sense that chemical forms come to be seen as fundamentally alchemical, "in which the pure gold of beauty and truth emerge purified by fire" (I/7, 294).

In the Munich speech, Schelling immediately draws out the consequences of Hamann's prophetic prognosis. To these murderers, nature is not merely mute, but it is *ein völlig totes Bild*—a consummately dead image, an image saturated with death (I/7, 294). This also suggests, therefore, a second sense of saturation, namely, an exhaustion or blockage in which form is a mere abstraction, ripped from its animating wellspring. Aldo Leopold, the American pioneer of the land ethic, makes Schelling's point in a much more contemporary, but also in a similarly prophetic, fashion.[9] When he articulates the "land ethic" and the awakening of an ecological conscience in which land is no longer reduced to the efficacious disposal of private property and "economic self-interest" (SCA, 209), he draws on an event in which his perspective on the land, which he had earlier more or less regarded as something with no value beyond its value to humans, is shattered by the sudden—miraculous even—coming forth of the fire of life.

In the backcountry of the Southwest, Leopold and his companions were eating lunch when they saw an animal that they first mistook for a doe. As it came closer, they realized that it was a wolf with a half dozen grown pups. "In those days we had never heard of passing up a chance to kill a wolf. In a second we were pumping lead into the pack." After they emptied their guns, they made their way down to the wolf.

> We reached the old wolf in time to watch a fierce green fire dying in her eyes. I realized then, and have known ever since, that there was something new to me in those eyes—something known only to her and to the mountain. I was young then, and full of trigger-itch; I thought that because fewer wolves meant more deer, that no wolves would mean hunters' paradise. But after seeing the green fire die, I sensed that neither the wolf nor the mountain agreed with such a view. (SCA, 130)

In his trigger happy youth, Leopold considered that the value of the land derived from its value to us as a place from where we can derive our sustenance and pleasures, just as Schelling characterized the view of nature as *Boden*, as mere "ground from which one draws nourishment and sustenance" (I/7, 293). For Leopold, the earth came to be regarded as mere property, to be disposed of in fashions efficacious to human interests. Because we live on the earth, we can take it for granted, that is, as our property, as something that, as given, we conclude that we own. Such a position, I would suggest in passing, cannot be separated from Schelling's account of radical evil, which "rests on a positive perversity [*Verkehrtheit*] or reversal of principles" (I/7, 367). One lives from the periphery as if from the living center, holding on to dead forms, including, most importantly, the *imago* of oneself, as if they were the wellsprings of life.

The death of nature in its relegation to property is nothing new, and Leopold also characterizes this relationship as Abrahamic (SCA, 204–205), but it has taken on an increased order of magnitude in the past three centuries. In the reduction of natural science to positivism, an equation against which Schelling always combated, the land has no value that in itself contests the exclusive gauging of its value in relationship to human interests. One need only think of the naturalistic fallacy and its refusal to derive ethical claims from nature. When nature is saturated with death, the value of its forms are assigned by those who regard it as something exclusively at their disposal. Land is valuable when it serves human interests and a wasteland when it does not. As Holmes Rolston III subsequently argued, "There is something overspecialized about an ethics, held by the dominant class of *Homo sapiens*, that regards the welfare of only one of several million species as an object and beneficiary of duty. We need an interspecific ethics. Whatever ought to be in culture, this biological world that *is* also *ought to be*; we must argue from the natural to the moral."[10] For Schelling, nature and art do not express something good, a particular thing or two that we esteem, but rather the progressive life of the Good itself. The soul "is not good, but rather it is the Good" (I/7, 312).[11] Without such an intuition, natural and artistic form express emptiness and "inner nullity [*Nichtigkeit*]" for they are "without the saturation of content" (*ohne die Fülle des Inhaltes*) (I/7, 305).

Leopold called the reversal of this perspective, the intuition of what Schelling called "infinite content" (1/7, 305), learning to "think like a mountain" (SCA, 129–133). When Leopold looked into the dying wolf's eyes, he did not just "see" that they were green but he "felt" their fire, the life that moved Leopold to think not from the *form* of the wolf but from the *ground* of the wolf. One must, to use Schelling's oft employed formulation, come to know the wolf so intimately that one moves *über x hinaus*, through *x* and thereby beyond *x*. As Schelling articulated this in the Munich speech: "We must go through the form [*über die Form hinausgehen*] in order to gain it back as intelligible, alive, and as truly felt [*empfunden*]" (I/7, 299). *Empfindung*, sensibility, is a difficult and critical term. We return to it in some detail in the next chapter, but for now we could at least say that it is the intuition of the life, the living depths, of the form, in the concrete apprehension of the form, much in the way that the Zen tradition speaks of the soundless sound and the formless form. Such an intuition, the shock of *Empfindung* in the sudden experience of the green fire, changes everything. When Leopold examined the wolf-free landscape, overrun by deer, he expected it to be a hunter's paradise, but instead found the flora of the mountains decimated by the exploding deer population. "Such a mountain looks as if someone had given God new pruning shears, and forbidden Him all other exercise" (SCA, 130–132). The awful majesty becomes the kitsch of an earth reduced to human economic interests. In the *Kunstphilosophie* lectures, Schelling strikingly claims that God is the immediate cause of all art (§23; I/5, 386), but, to borrow the phrase from the *Freedom* essay, God cannot be a "God of the living" if nature is, as it is in radical evil, saturated with death. In evil, we are all dancing with a good conscience and intellectual self-certainty in a great self-enclosed circle of pure light. In our dance, we are not *like* God but *as* God. Such a dance in the realm of detached light is the fire not of life but of hell, as if the relentless proliferation of suburbs and shopping malls expressed the grace of nature.

I think that for this reason the great Zen Master Eihei Dōgen also loved mountains and counseled us to think like mountains, whose dominating forms display not only the presence of form, but also, in their emptiness (lack of intrinsic, free standing identity), the saturation of Dharma. "But after entering the mountains, not a single person meets another. There is just the activity of mountains. There is no trace of anyone having entered the mountains."[12] Mountains are not property. "Although the mountains belong to the nation, mountains [really] belong to people who love them. When mountains love their master, such a virtuous sage or wise person enters the mountains" (S, 162). Hence Dōgen tells us that we should "know for a fact that mountains are fond of wise people and sages" (S, 163). Thinking like a mountain demands that we "do not view mountains from the standard of human thought" (S, 163).

If art imitates nature, then not only must we rethink nature by learning to think like a mountain, but we must therefore also reconsider what it means to imitate. Starting with nature itself, how does nature repeat again and again its own progression? How does nature imitate nature? As we also saw in the third chapter, nature does not progressively imitate itself by merely repeating its forms recursively as if they were laws. In the 1800 *System*,[13] Schelling speaks of the "*free* μίμησις [*freie Nachahmung*]" of the act of self-consciousness, the act by which eternity again and again becomes transposed as living form, as that "with which all philosophy begins" (ST, 65). Indeed, "philosophy in general is nothing but free μίμησις, a free repetition [*Wiederholung*] of the original series of actions in which an act of self-consciousness evolves itself" (ST, 66).[14]

In the Munich speech, Schelling contrasts "servile μίμησις [*dienstbare Nachahmung*]" (I/7, 294) and its "tangible lack of life" (I/7, 300) with "vital μίμησις" (I/7, 301). Servile imitation, which reproduces and represents forms as if the forms themselves were stillborn, allies itself with the nature-cide of positivism. "Death and unbearable severity would be the art that wanted to present the empty husk or delimitation of the individual" (I/7, 304). As Kandinsky famously claimed in 1911, "An effort to animate past principles of art, at best, results in artworks akin to stillborn children."[15] Living imitation, moreover, has the force of a miracle, "the miracle [*das Wunder*] by which the conditioned is elevated to the unconditioned" (I/7, 296); it is the beauty that grips us "with the power of a miracle [*mit der Macht eines Wunders*]" (I/7, 315). As Georges Bataille once characterized the miraculous: impossible, but there it is![16] A future that could not have followed from what preceded it nonetheless, unexpectedly and unprethinkably, comes forth. The miracle is the temporality of nature and art, its living temporality. As William Blake articulated it in one of his wonderful "Proverbs of Hell" from *The Marriage of Heaven and Hell*: "Eternity is in love with the productions of time." This love in the free and living repetition of its temporality is the grace (χάρις, *Anmut*) of the soul as it expresses itself in the green fire of form.

Finally I return to the masks themselves, whose saturated progress expresses the marvelous nonsequiturs of free, living μίμησις. The carnival is the festival of plasticity itself, of *das Bildende* as such. How do we think of plasticity as living masks, symbols, and husks? As a first hint as to how one could here this word, we can look at how in her recent work, *Plasticity at the Dusk of Writing: Dialectic, Destruction, Deconstruction*,[17] Catherine Malabou understands the "conceptual portrait" of her own work as a *transformational mask*. Found in several cultures including the Pacific Northwest, a transformational mask is a mask that opens up to reveal another mask beneath it.[18]

Such a mask is, however, no "mere mask," no hollow husk, no lifeless symbol. The mask for Malabou, following Claude Lévi-Strauss,

Kwakiutl (Kwak'wala) transformational mask by John Livingston (see note 18)

indicates "an agonism between form and its dislocation, between systematic unity and the explosion of the system" (PDW, 6).

> Transformational masks never reveal the face they mask. They are ill-suited to the human face and never marry the model, nor are they designed to hide it. They simply open and close onto other masks, without effecting the metamorphosis of someone or something . . . rather than disguising a face, the masks reveal the secret connection between *formal unity* and *articulation*, between the *completeness of form* and the *possibility of its dislocation*. (PDW, 2)

Transformational masks express the living temporality of plasticity. "Plasticity thus appeared to me from the outset as a *structure of transformation and destruction of presence and the present*" (PDW, 9). The carnival is the temporal progression of masks, endlessly unfolding layers of divine personae without God being something beneath the masks, but rather the dynamically productive excess of their visibility, the dark mother of gravity.[19]

We could now say that just as nature is not only *naturata* (the clarity of what is), but also the productivity of *natura naturans* (nature producing itself anew), art is not just the catalog of artworks and techniques, but also the miracle of art. Nature natures and art arts!

At the beginning of the Munich speech, Schelling argues that the "relationship" between nature and the *Bilder* of art, the plasticity or shaping into form brought forth in *die bildende Kunst*, is located in a "living center" that holds art and nature together. "The plastic arts therefore stands manifestly as an active copula [*Band*] between the soul and nature and can only be grasped in the living center between both of them" (I/7, 292). What is this living center that governs the relationship between φύσις and τέχνη, nature and art? How do the images of nature relate to the images of art? *The living center is the imagination itself.* As we see in greater detail in the following chapter, the life of *die Bilder* is the productivity of *Einbildungskraft*.

The latter already literally speaks of the sovereign introduction (*ein*) of image (*Bild*) into that which is at first without image. Schelling sometimes spoke of this "expulsive [*ausstoßende*]" movement as the *In-Eins-Bildung* (e.g., I/7, 60), that is, the conjunction of soul and form as a saturated production. The many become one (*Eins*) through have coming into (*In*) form (*Bild*). Coleridge, for his part, attempted to render this movement through his remarkable neologism "esemplastic," derived from the Greek "εις εν πλάττειν, that is, to shape into one. . . ." The shaping is "plastic" (from πλάττειν), suggesting the unified and unifying movement from the formless to the formed. As Coleridge reflected: "I constructed it [the word 'esemplastic'] myself from the Greek words, εις εν πλάττειν, i.e., to shape into one; because, having to convey a new sense, I thought that a new term would both aid the recollection of my meaning, and prevent its being confounded with the usual import of the word, imagination."[20] The relationship between the plastic arts and nature is the plasticity of the esemplastic itself.

When I repeat the movement of the imagination as if I were simply reproducing something I once saw as if it were secondhand version of the same thing, art is lost and representation prevails. (It was the genius of Proust to demonstrate that remembering is creative.) Kitsch, to use a term made famous by Hermann Broch, has a murderous relationship to nature. Kitsch is not, as Broch insisted, bad art, but rather pseudo-art, something nonartistic trying to pass itself off as art. As such, Broch concluded, kitsch is an experience of radical evil. "The maker of kitsch does not create inferior art, he is not an incompetent or a bungler, he cannot be evaluated by esthetic standards; rather, he is ethically depraved, a criminal willing radical evil."[21] Schelling already implied this in his introduction to the *Philosophy of Art*. The older and genuine artists "went from the center to the periphery" (I/5, 360). In the *Freedom* essay, we learn that evil is to abandon the center and from the periphery make oneself the center (we return to this in our discussion of Captain Ahab in the next chapter). Kitsch takes the dead forms of art without a sensibility for the living ungrounding ground of art. It is a cheap and dead imitation in which "the later artists take the external and extracted form and seek immediately to imitate them. They retain the shadow without the body" (I/5, 360). Kitsch operates in the shadows.

The life of the imagination is always a struggle and sometimes our artworks are just not all that good, but failed art is still art. The struggle for art is always also the ongoing struggle against the lurking forces of kitsch. In Milan Kundera's novels, which eschew the facile world of stereotypes and economies of imitation, kitsch is the "absence of shit" in the sense that it is a rejection of all that does not accord with itself: "Kitsch has its source in the categorical agreement with being."[22] Schelling had long understood this agreement as the danger of dogmatism, which we can now see is inseparable from the problem of evil. This can readily be seen in the catastrophic ascendency of politicians who have no sensibility or exposure to art (I/5, 352). The reign of Goethe's philistine and Schelling's *Verstandesmensch* is apocalyptic.

We could also say, extending Schelling's position, that kitsch is a kind of *Doppelgänger* of art, haunting us from the realm of the dead. In this saturating of death, we see that we have more fundamentally lost faith in life. Art calls us back, paradoxically, beyond the mistake of positing an exclusive disjunction between art and nature, to a more *natural and creative* attunement to life. As Deleuze argued in his text about cinematic images of time:

> Cinema seems wholly within Nietzsche's formula: "How we are still pious." . . . The modern fact is that we no longer believe in the world. We do not even believe in the events that happen to us, love, death, as if they only half concerned us. It is not we who make cinema; it is the world that looks to us like a bad film. . . . The link between the human and the world is broken. Henceforth, this link must become an object of belief: it is the impossible which can only be restored within a faith. . . . The cinema must film, not the world, but belief in this world, our only link. . . . Whether we are Christians or atheists, in our universal schizophrenia, *we need reasons to believe in this world.*[23]

When we believe in the world, when the study of science inspires our love of poetry and when the love of poetry awakens our passion for science, even the past is reimagined as the unprethinkable future. Belief in the world—what Schelling might call *faith in the earth*—exposes and disables our clichés of living and dying, and in the relationship between art and nature, we become vulnerable to the green fire that is the shock of real tears and the trembling of real laughter.

The call in the 1797 *System* fragment for "monotheism of reason and the heart, polytheism of the imagination [*Einbildingskraft*] and art"[24] can also serve as one of the great arguments for the intrinsic value of biodiversity. Hearkening back to our opening chapter, Schelling could not have fully anticipated the current catastrophe of the Sixth Great Extinction event. Nonetheless, the very movement of his thinking illuminates the speciesist threat to biodiversity—that we largely as a species only accord value to ourselves and our interests and consequently threaten the earth with the ruinous monoculture of our species ego—as one of the greatest issues in the whole history of our species. It is an issue on which everything else that belongs to human freedom (*die damit zusammenhängenden Gegenstände*) depends.

The Schellingian gesture resonates in the twenty-first century as the following decisive question, a question that also resonates with Heidegger's *Beiträge*: Will our relationship to the monstrous depths of the earth have been on balance one that drove us to the periphery and in our anxiety will we have obsessively endeavored to remake the earth in our image—a destiny that not even we will survive? *Or* will this be the awakening to a new earth, a self-closing earth that bears a world without image,[25] an earth already prophesied in Samothrace (at least in Schelling's account of it)?

It is to this outrageous but decisive final possibility that we now turn.

6

Life of Imagination

Where on Earth

Schelling's 1809 *Freedom* essay is widely regarded as his masterpiece and rightly so. Even Hegel, with whom Schelling would become increasingly estranged, recognized in his 1825 to 1826 *Lectures on the History of Philosophy* that it was "deeper and more speculative."[1] Heidegger, despite rather ludicrously aligning Schelling with Nietzsche as the harbingers of the consummation of metaphysics in its turn to the subjectivity of the will, called it the "heat lightning of a new beginning."[2] In the *Freedom* essay, Schelling announces that he is continuing his efforts to develop more fully his philosophical project by no longer starting with the things of nature and culminating with freedom (as he had in the *Naturphilosophie*, which he came to understand as negative philosophy), but rather by exposing the reverse direction. How does philosophy, including the human and her earth, appear when it is not asked about from the periphery but from the center, from the ungrounding ground of freedom? What happens, we might say, when philosophy becomes more imaginative and not merely a detached account of the imagination?

The impact of this reversal—no longer thinking toward the center but from the center—gives the *Freedom* essay its great originality, but the latter does not in itself fully prepare us for works like the marvelous 1815 Samothrace address. From whence this invocation of the ancient gods, as if Schelling were attempting to emancipate the signs of their former vitality from their ossification in ancient public documents? Such thoughts do not come from elsewhere or from on high. "None of our spiritual thoughts goes beyond the earth" Schelling scolded Eschenmayer (I/8, 169). How then does the thought of freedom as the ground of the human take us to the far-flung isle of Samothrace and its ancient, proto-pagan cult of Cabiri?

In this 1815 address, subtitled a *Supplement* [*Beilage*] *to the Ages of the World*, we learn:

> The modest isle of Thracian Samos possessed something still more splendid, something with which it could enrich the history of humanity, to wit, the cult of the Cabiri. It is the oldest cult in all Greece, having come to light early on in the region, offering the island its matutinal aura [*mit dem das erste Licht höheren*] and a still more elevated form of knowledge. The cult did not disappear until the ancient faiths themselves foundered. From the forests of Samothrace ancient Greece first received, along with a more mystery-laden history of its gods, belief in a future life. According to universal ancient testimony, those who were initiated into the rites were better prepared to live and to die, and happier in both their living and their dying. Offering sanctuary to victims of misfortune, and indeed to criminals—insofar as reconciliation could be achieved through confession and remission of sins—the rites of Samothrace preserved a feeling for humanity in both archaic and more recent times of savagery. No wonder the name of this holy island came to be interwoven in the most ancient stories touching everything of honor and renown.[3]

This wondrous gift of ancient history whose rituals helped one become "better prepared to live and to die" and made one "happier in both their living and their dying" revealed that philosophy could have practices of awakening and self-transformation such that the philosopher could become a different kind of person, one, for example, generous enough to forgive crime and hurt.[4] That philosophy would awaken, just as Nietzsche in his own way later saw, to ways of living and dying that were not dominated by *ressentiment*, anxiety, and small-mindedness, suggests that the task of philosophy is not only to analyze and narrow in on issues, but also to cultivate the largesse and creativity that the imagination unleashes. Schelling was not just a philosopher of creativity and the imagination. He was a creative and imaginative philosopher.

Such creativity has nothing to do with being willful. It is rather *the composure that lets be* in the sense of *Gelassenheit*. We associate this term with Heidegger although it was Schelling, taking the term from Meister Eckhart, who first deployed it as a way out of the tacit willful and egoistic subjectivity that characterized modernity. That Heidegger would have accused Schelling himself of being guilty of this is itself ironically an exceedingly willful reading. For Schelling, thinking, living, and dying, all of these come to life with *das Wirkenlassen des Grundes*, letting the ground operate (I/7, 375). This has nothing to do with willing but is, as Schelling felicitously articulates it in the 1815 *Ages of the World*, "the will that wills nothing" (I/8, 235). Willful subjectivity has nothing to do with creativity and the imagination. It is, rather, the human experience of evil, perversely assuming the center into oneself. Only in letting go of oneself and letting the ground operate does the imagination show itself both in human works and as the wild way of nature itself.[5] It is also at the heart of what Schelling will call philosophical religion, the creative—historic yet prophetic—reimagination of the earth and the humans who dwell, often ruinously, on it.

At the heart of both this chapter and this book, then, is this great problem of *die Einbildungskraft*, the imagination. But what is the imagination and what is its fruit, that is, *what is an image* or *Bild*? What exactly is the *Bild* in *Einbildungskraft*? At the heart of this problem one will find some of Schelling's most daring and extraordinary thinking, thinking that gives him an impressive interdisciplinary range and whose prophetic force makes him our contemporary.

Just as the vital promise of Samothrace demands its genealogical excavation from the texts and traditions that enervated it, the retrieval of the problem of the image also demands that we disinter and reclaim it from the ravages of Platonism in order to rethink it in the multidisciplinary context of Schelling's concern with freedom.

Both Senses of Sense

Deleuze argued in *Difference and Repetition* that "the task of modern philosophy has been defined: to overturn Platonism [*renversement du platonisme*]" (DR, 59/82). Even for Deleuze this was not the demand that we simply get beyond Plato—"the Heraclitean world still growls in Platonism" (DR, 59/83)—but it includes moving beyond the stark dualism of the intelligible realm of being and the sensible one of becoming. I belong to those who, like Schelling, do not hold Plato accountable for the immensely and perniciously impactful *Wirkungsgeschichte* of this decisive distinction. Plato was already post-Platonist *avant la lettre*, and the two-realm doctrine was perhaps an unhealthy legacy of the worst part of Christian metaphysics—Nietzsche's "Platonism for the people"[6]—in which, exhausted with the vicissitudes of becoming, we sought comfort in the fantasy of the permanence of a detached intelligible realm. Nonetheless, even in an increasingly secular, post-Enlightenment context, this fateful assumption, even as God drops out of the equation, still tacitly influences how we think about the relationship between sensibility and intelligibility, especially when it comes to art and religion. Schelling had struggled from early on with this task of modern philosophy, a task that he also associated with what he considered constitutive of modern philosophy, which, as we have seen, has "since its inception (in Descartes) the shared deficiency that nature is not present to it and that it lacks nature's living ground" (I/7, 356–357).

This reclamation of the inseparability of both senses of sense (intelligibility and sensibility) within sense itself was already hinted at in the 1797 *System* fragment:

> At the same time we so often hear that the great multitude should have a sensible religion. Not only the great multitude, but even philosophy needs it. Monotheism of reason and the heart, polytheism of the imagination [*Einbildingskraft*] and art, that is what we need![7]

Regardless of the silly scholarly fight since Rosenzweig over the identity of the author of this fragment,[8] it is clear that this intuition shapes Schelling's entire path of thinking. The call for a *sinnliche Religion* is not the call to abandon rationality for the vagaries of a *Naturschwärmerei*—this is one of the fateful steps that eventually led to the extreme reactivity of later German Romanticism. One need only think of Klages' "rhythm of cosmic life" whose loss led to a "servitude of a life under the yoke of concepts. . . . To again free life from this, the soul as well as the body, is the hidden draw for all mystics and users of narcotics."[9]

Sinnliche Religion is not the wanton and nostalgic abandonment of intelligible sense for the other sense of sense. The adjective *sinnlich* derives from *Sinn*, which, like the English word "sense," preserves sense in both senses of sense. Overcoming Platonism demands neither that we locate the intelligible sense of the sensuous in some remote realm, wholly otherwise than and ontologically distinct from the sensuous (the reality beyond appearance), nor that we abandon the yoke of the intelligible for the allegedly senseless realm of the sensuous.[10]

Moreover, this sensuous religion, at both levels of sense ("reason and the heart") is at the level of its ungrounding ground *one* (monotheism), but that does not make it mathematically one, for it can in no way be thought of as any one thing or being. It is the *one* that can only be thought of as *many*, never as *einerlei* (any one single thing). It is the *one*, so to speak, of *difference as such*.[11] The *one* is the chaos that in its *vulnerability* or *weakness*, to use the language of Schelling's later Philosophy of Revelation, can become anything, but which in itself is nothing, much as the Mahāyāna tradition speaks of emptiness or *śūnyatā* as water, which can take any form because it has no form of its own. Except as an abstraction, it is not found in itself, but only as the univocity of being:

> A single and same voice for the whole thousand-voiced multiple, a single and same Ocean for all the drops, a single clamor of Being for all beings [*une seule clameur de l'Être pour tous les étants*]: on condition that each being, each drop and each voice has reached the state of excess—in other words, the difference which displaces and disguises them and, in turning upon its mobile cusp, causes them to return. (DR, 304/388–389)

It is the one that can never express itself as *one*, but only as difference, as that which has always not happened in the singularity of anything that has happened. (Its singularity already insists that it does not mean anything beyond itself.) This *one* that negative philosophy presents (*darstellt*) is inseparably and simultaneously the "polytheism of the imagination [*Einbildingskraft*] and art." God is always gods and art is always arts. As Jean-Luc Nancy articulates this subtle but critical distinction, it is not the "principle *of* plurality" but rather the "plural itself as principle,"[12] the being-singular-plural of the gods and arts.

"Multiplicity exposes unity in multiple ways" (M, 31) and in so doing reveals the world as

> simply *patent*, if one may understand by that an appearance that does not "appear," no immanence of a subject having preceded its transcendence, and no obscure ground its luminosity . . . on the one hand, the appearing of the nonapparent, or the nonappearing of all "patency," and on the other hand, this, that there is only the world. (M, 33)

This is at the heart of Schelling's thought of the imagination or *Einbildungskraft*: it can be itself only by always being all others, each in its singularity, each meaning only itself. The singularity of artworks, each speaking from and for itself, is obscured when they are merely regarded as examples of the idea of art as such. Resisting Hegel, we can never be finished with art because art is not merely a sensuous absolute whose intelligible idea can now be discerned. Hegel argued that art, which once upon a time presented "the depths of a *supersensuous world*," no longer speaks to our deepest interests. The time of the sensuous presentation of the supersensuous *as the supreme moment* of such presentations is over. As he famously proclaimed at the beginning of his lectures on aesthetics:

> The spirit of our world today, or more particularly, of our religion and the development of our reason, appears as beyond the stage at which art is the supreme mode of our knowledge of the Absolute. The peculiar nature of artistic production and of works of art no longer fills our highest need. We have got beyond venerating works of art as divine and worshiping them. The impression they make is of a more reflective kind, and what they arouse in us needs a higher touchstone and a different test. *Thought and reflection have spread their wings above fine arts*. (italics mine)[13]

That thought and reflection could spread their wings above fine arts, indeed, that sense and sensibility are separable and, by implication, that sensible works do not in their own way think, suggests a nascent and subtle form of Platonism. John Sallis, in his *Transfigurements: On the True Sense of Art*, decisively detects the subtle play of the bifurcation of the intelligible and the sensible in Hegel's claim that art has exhausted itself, giving way to a more direct revelation of the thought and reflection implicit in them:

> If, in the course of the development of truth, it comes to surpass the possibility of being adequately presented in a sensible configuration, if, beyond a certain point in its development, it proves to exceed every possible sensible form, then this point will mark the limit of art and will delimit the past to which art, even in the future, will remain consigned. In a certain respect this limit is posited from the moment truth is comprehended as the intelligible over the sensible.[14]

Despite his remarkable and exhaustingly thorough analysis of art, for Hegel art works do not come simply as themselves but are always in the end about something other than themselves (a sensuous presentation of the intelligible idea of the absolute). For Schelling in contrast, *art is not merely the idea of art*, but the singularity of works and, as we shall see, art cannot be separated from religion and hence it cannot be separated from mythology because artworks, like the gods, only come as themselves. Art works only mean themselves and do not mean something remote to themselves, and nonetheless they are the living presence of an unpresentable life.

For the life of imagination to present itself in the living singularities of its works ("polytheism"), it needs something so rare that it appears unique to the *System* fragment authors.

> First I will speak about an idea here, which as far as I know, has never occurred to anyone's mind—we must have a new mythology; this mythology must, however, stand in the service of ideas, it must become a mythology of *reason.* [*Vernunft*]

This is the call for what Schelling will later call *philosophical religion*, an intimation of what in the *System* fragment is the inkling that "a higher spirit sent from heaven must found this new religion among us. It will be the ultimate and greatest work of humanity."[15] This strange demand insists that "philosophy must become mythological in order to make philosophy sensible [*sinnlich*]." How are we to think *the becoming sensible of philosophy*, including its relationship to art, religion, and history, when the sense of philosophy's sense has always been the intelligible over and against the aesthetic and sensuous (Platonism)?

The War of Monoculture Against Monoculture

When Kant attempted in 1798 in *Der Streit der Fakultäten* to overcome the bellicose conflict between the competing disciplines (a war over what is mine and yours) and to vindicate the rights of philosophy to think for itself and not be censored, we can see an early version of what is now an old problem: academic work is a territorial property war.

For Schelling, it does not have to be this way. Just as the Cabiri initiates were "were better prepared to live and to die, and happier in both their living and their dying," Schelling notes in the second of his introductory aphorisms to the *Naturphilosophie* that:

> Wherever that revelation [i.e., the divinity of the All] came to pass, even if it was merely transitory, there was always inspiration [*Begeisterung*], the discarding of finite forms, the cessation of all conflict, and unity and wondrous consensus, which often cut across long eras. It was accompanied by the greatest spiritual characteristics whose fruit was the general alliance between the arts and sciences. (I/7, 140)

This mysterious thought, the secret treasure at the heart of the imagination, produced, perhaps incredibly, enduring periods of peace whose precise mark was the absence of the conflict of the faculties, the cessation of the academic turf wars over what is mine and yours, and a seemingly miraculous feeling of cooperation that did not sacrifice the autonomy of the various disciplines, but rather allied them. To be clear: fundamentally at stake in Schelling's entire philosophical path of thinking is a transformative experience of philosophy—and of nature and science and art and religion and history—that is radically affirmative and loving and thereby *gelassen*, the stance of *Zulassung* (I/7, 375),[16] an "allowance" that lets nature become (progress, create) from *out of a primordial ocean of peace*. To be at peace with all things, even difficult and painful things—this is the joy of religious life and the inspiration for philosophical investigation. We can scarcely imagine this rare disposition of composure and affirmation. It seems historically especially remote to institutional religions and certainly to academics. As Schelling laments in the subsequent aphorism:

> Whenever the light of that revelation faded and humans knew things not from the All, but from other things, not from their unity but from their separation, and in the same fashion wanted to conceive themselves in isolation [*Vereinzelung*] and segregation from the All, you see science desolated amid broad spaces. With great effort, a small amount of progress is made, grain of sand by grain of sand, in constructing the universe. You see at the same time the beauty of life disappear, and the diffusion of a wild war of opinions about the primary and most important things, as everything falls apart into isolated details [*alles in Einzelheit zerfallen*]. (I/7, 140)

In the *Freedom* essay, Schelling expressly writes against the *zu oft beherrschender Sektengeist* (the too often prevailing sectarian spirit) that plagues German thinking, seeking rather *die freie kunstreiche Ausbildung*, the education or formation that is ingenious and rich in art, rather than the prevailing *künstliche Schraubengänge der Polemik*, the artificial screw-threads of polemic. In the *Philosophy of Art*, Schelling speaks against the literary Peasant Wars. Thinking after the death of God, that is, thinking after the loss of faith in any discernible and generally orienting first principle, academic freedom resembles Hobbes war of all against all as difference becomes a liability in "the diffusion of a wild war of opinions about the primary and most important things." A peaceful unity, moreover, cannot be imposed from without—it "is not a light that simply illumines from without, but rather one that excites from within" (I/7, 141). It is realized by abandoning the periphery altogether and returning to the center where "not only are the sundering of the sciences from one another mere abstractions" but "so is the sundering of science itself from religion and art" (I/7, 141).

Although Schelling often speaks in terms of the center, including his dramatic evocation of it in the *Freedom* essay, how are we to think it and how does it manifest in the cooperative being singular plural of the disciplines? In the 1806 essay on gravity and light, we find a remarkable sentence: "The light essence is the gleam of life in the omnipresent center of nature [*Das Lichtwesen ist der Lebensblick im allgegenwärtigen Centro der Natur*]" (I/2, 369). *Der Lebensblick* can also be translated as the glance, look, or sight of life. It is gravity becoming sensible, coming into image, into a look, into a shape or form, as in εἶδος. It is the look, so to speak, of what in itself is visible precisely as invisible. It is the paradoxical look of monstrosity.

Hence the equally remarkable preceding sentence: "Only the darkness of gravity and the gleam of the light essence together produce the beautiful shining of life [*den schönen Schein des Lebens*], and turn the thing into the properly real, as we call it" (I/2, 369).[17] The unity of the potencies of gravity and light, the living copula (*das Band*), the A^3, is *die schöne Schein des Lebens*, life shining forth in the beauty of shining as such, the image as the revelation without representation of gravity. This *Schein*, this *Vorstellung* that is the coming forth, the positing as time, of the center, allows us to see that we see nothing of the center, not even that in any proper sense it is even a center. (It is closer to how Badiou understands the empty set, "emblem of the void, zero affected by the barring of sense."[18]) There is only the patency of *die Schein* itself, the presentation of the earth in its contracting gravity as image (*Bild*), the life of *Einbildungskraft* (the coming to light of what in itself never comes to light). At the beginning of *Moby-Dick*, Ishmael, with his journey to the sea looming as death loomed for Cato—"with a philosophical flourish Cato throws himself upon his sword; I quietly take to the ship"—makes the same point about the outlandish ocean: "the image of the ungraspable phantom of life; and this is the key to it all" (chapter 1, Loomings).

Art of Imagination

While the *Naturphilosophie* as negative philosophy allows phenomena to emerge in an intellectual intuition as an expression of the ungrounding of their ground, we already see in the 1800 *System of Transcendental Idealism* that art and aesthetic intuition do something that philosophy, simply left to its capacity for reflection and critique, cannot supply for itself. As we saw in the opening chapter, reflection must first return from the periphery to which it first withdrew as it breaks the grip of the seeming iron cage of necessity that it had initially identified with nature. It broke from the natural attitude and the spell of presence, so to speak, simply by virtue of one's perceptions becoming questionable. When Plato's prisoner turns around and sees that seeing itself has been rigged, the perceptual realm immediately reveals itself to be an image, although the prisoner cannot at this point say of what the image is an image. All the prisoner can say is that the image is as such because one can see that one does not know what one is seeing (see chapter 1).[19] One no longer sees pictures that exemplify remote and familiar meanings (perception as the recognition of the sensuous exemplification of intelligible meanings). The image first looms as an image because one does not know what to make of what one sees. The progression of *Naturphilosophie* is from the periphery where one is the subject and images appear as remote and mysterious objects, back to the center, where the subjectless subject is nature itself as the sovereign life of the imagination, of *natura naturans*. It moves from the real of *natura naturata* to the ideal temporal self-positing or autogenesis of *natura naturans*. Reflection is born from the cracking of the natural attitude, but it completes itself by returning to the heart of nature no longer as necessity, but now as sovereign.

The production of works of art, the unprethinkable coming into form of the in-itself formless, is the reverse progression, the *Darstellung* or presentation of the ideal as real. The creation of artworks, we learned already in the 1800 *System*, is not in any way contingent on the being aware through philosophical reflection of the structure of creation. It need only be a "productive intuition": "The fundamental character of the artwork is therefore an *unconscious infinity* [bewußtlose Unendlichkeit]" (STI, 290; §2 Character of the Art Product 1). The artist does not have to appreciate intellectually that creativity is the schematic imagination of infinity in order to create. Art does not need philosophy to produce art works, although philosophy arguably can be of great benefit to it. Philosophy, however, needs art. Otherwise it remains simply negative, lacking the progression from infinity, that is, lacking its own inherent creativity and life. "For the aesthetic intuition is precisely the intellectual intuition become objective" (STI 296; §3 Corollaries, 1). Creativity is not some subjective fancy, but rather the intellectual intuition reversing its emphasis and becoming productive, *creating from prime*—or more precisely, affirming the prime in the unfolding of its creativity. It expresses in an objective form the "unconscious [*das Bewußtlose*] in acting and producing" (STI, 299). Art opens up for philosophy the "holy of holies" (STI, 299)—the original unity of that which was first severed when the natural attitude sundered nature from history. Without this break, on the one hand, there can be no real history because nature is assumed to be the recursion of necessities. Without art, philosophy does not return from the periphery of reflection and therefore regards nature as distinct objects for a discerning subject.

One can already hear the faint echo of the call for a new mythology in the return of science to poetry in which the "consummation of philosophy flows like individual steams back into the ocean of poesy" (STI, 300), which reciprocally, as we shortly see, allows the ideas of nature to appear objectively, that is, as the real. Moreover, what are we to make of this strange and monstrous image of the ocean that Schelling calls a primordial image or an *Urbild*, itself defying representation as a picture and whose concept always leaves an irreducible remainder, yet which emerges as the unimaginable and unrepresentable source of images?

In a September 3, 1802, letter to August Schlegel, Schelling explains that his lectures on the philosophy of art will attempt to reflect critically on "art in itself, of which the empirical only provides the appearance [*Kunst an sich, von der die empirische nur die Erscheinung gib*t]." What is art such that appearance (*Erscheinung*), shining forth, is but its product? This question demands that Schelling speak "of the root of art, how it [art] is in the absolute . . . art therefore taken entirely and simply from its mystical side."[20]

In the fourteenth of his 1802 *Lectures on the Methods of Academic Study*, Schelling indicated that the philosophical critique of art has nothing to do with the study of its formal techniques or even a history of art's formal accomplishments; we are studying "a more sacred art, which the ancients called a tool of the gods, a bearer of divine mysteries, the revealeress of the ideas [*ein Wekzeug der Götter, eine Verkündigerin göttlicher Gehemnisse, die Enthüllerin der Ideen*]" (I/5, 345). Art not only presents works, but it does so in a way that is both revelatory of presentation as such and a testimony to the unprethinkable life of the imagination, much as Wallace Stevens was later to speak of the imagination as "a moment of victory over the incredible" and "the power that enables us to perceive the normal in the abnormal, the opposite of chaos in chaos."[21] Poetry, counter to philosophy's official account of what we imagine to be our world—reflection contracts while the imagination expands—, renegotiates the boundaries of the true and the credible—an "unofficial view of being" (NA, 667).

Hence, Schelling defended Plato's infamous banishment of the tragic poets in Book 10 of the *Republic* not because philosophy lords the idea over the image, but rather because these poets were merely "realistic," that is, they provided pictures or representations that they uncritically passed off as the real rather than *producing images as the real*.[22] This banishment was not the inaugural act of Platonism in which sensible images mean an idea remote to them, but rather the liberation of the imagination from the fantasy that it is a faculty of representation. It is not any truer to say that the imagination is the fanciful contrivances of the human subject. Art is not the arbitrary and willful wandering of the will, but rather precisely demands its mortification and abnegation because art is "incomprehensible [*unbegreiflich*] in its origin" and "miraculous in its effect" (I/5, 347). Its products do not follow from a discernible first principle; artworks cannot be deduced but rather emerge from the *geheime Urquelle*, from the enigmatic primordial source. Art is the consummate coming into image of the real from the ideal (die *vollkommene In-eins-bildung des Realen und Idealen*) (I/5, 348).

Art is not the illustration of a priori ideas, the sensuous representation of intelligible content. In his 1812 response to Eschenmayer (see Appendix A), Schelling takes up the critical but vexing relationship in Plato between sensuous beings and their primordial forms:

> It is also held by true Platonists. For Plato says for example in the *Phaedrus* of *beauty*, that we see it first in the highest glance and recall it when we here view beautiful things. To the extent I can presume to understand Plato's meaning, he means here the original beauty—not what-has-being-for-itself [*für-sich-seyende*] since in itself it cannot be a subject, but the beauty *of what is in and from itself beautiful*, and this, as an actual *subject* (an ὄντος ὄν), can alone be called a *primordial image* [*Ur-Bild*], an *idea* according to the Platonic use of language and manner of thinking. It is also well known that Plato does not speak of *equality*, of *beauty*, *goodness*, *justice*, *saintliness*, but of the *equal itself* (αὐτὸ τὸ ἴσον, *Phaedo*, 74a),[15] of *beauty*, of *goodness*, of *justice*, of *saintliness* themselves (περὶ αὐτοῦ τοῦ καλοῦ, καὶ αὐτοῦ τοῦ ἀγαθοῦ καὶ δικαίου υκαί ὁσίου, *Phaedo*, 75d). Now it is frankly true that he who sees the good in itself, the beautiful from itself, the equal in itself to itself, also views the primordial image of *beauty*, the primordial image of *goodness* and *equality*. But the primordial image of these things is not beauty as a property or as an abstraction, but rather what has being [*das*

> *Seyende*] (τὸ ὄν), the subject, primordial beauty [*das Ur-Schöne*] itself. (I/8, 179–180)

Beauty does not have a separate existence from beautiful things and is only given in and as beautiful things. It has not been lying around from time eternal (a nonsensical idea in itself according to Schelling), waiting to be instantiated. Beauty is in the image as an image and comes to us in a primordial way through and as images.

Ideas are not freestanding a priori fixed rules that play themselves out recursively in the sensuous realm. Beauty and goodness, immanent within sense itself, are themselves creative, progressive, and evolving. In art this comes to be associated with genius. In the *Methods of Academic Studies* lectures, Schelling argues that we can discern philosophically what operates unconsciously in genius, that is, philosophy can "know (*erkennen*)" the "autonomy" or "absolute legislation [*Gesetzgebung*] of the genius" (I/5, 349). Kant in the first *Kritik* had argued that the orderly and necessary realm of representation (*Vorstellung*), formed according to space and time as the pure forms of intuition, was the consequence of the transcendental legislation of nature. For Schelling, the genius demonstrates that this legislation is *sovereign*. This is not to say, however, that the genius is a sovereign *subject*. Creativity is not the result of the caprices and eccentricities of the artist-subject—freedom has the human; the human does not "have" freedom. Genius is rather "animated [*beseelt*]," *simultaneously* sovereign and necessitated, creating "in a freedom akin to the gods and at the same time the purest and highest necessity [*in einer gottähnlichen Freiheit zugleich die reinste und höchste Notewendigkeit*]" (I/5, 349). It is the sovereign presentation of form.

In the *Philosophy of Art* lectures, we again find the operation of this simultaneous double movement. Genius in human activity is the operation of the δαίμων, "the indwelling divine within humans [*das inwohnende Göttliche des Menschen*]" and "a piece of the absoluteness of God" (§63, I/5, 460). The genius creates while desiring exclusively what is necessary; there is nothing feigned or affected. The genius *materializes* the ideas (the potencies) and gives them a life of their own. "God is the absolute cause of all art" (I/5, 386). One who is free for art is "at the same time passive and active, carried away and pondering [*zugleich leidenden und tätigen, fortgerissen und überlegten*]" (I/5, 358–359). In being torn away from the official view of being, one is not merely beside oneself in some frenzied thought-free zone. This transport is simultaneously a brooding and pondering, a thinking of what heretofore had been unthinkable. This brooding, however, is not the exercise of philosophical analysis, but rather the *wondrous advent of meaning as such*. Philosophy cannot "bestow meaning [*Sinn*] to art, only a god can bestow it" but it can open again the "primordial sources of art for reflection [*Urquellen der Kunst für die Reflexion*]" (I/5, 361), the "true primordial sources of art out of which form and matter [*Stoff*] flow undivided" (I/5, 360). In the dialogue *Bruno*, written around the same time as the art lectures, Schelling argues that philosophy as such is esoteric, that it seeks to articulate and speak to what is entailed in manifestation. The esoteric movement of thinking does not produce the beautiful, but it reflects on the idea of the beautiful. Conversely, art is exoteric. The concrete beautiful object is the beautiful shining forth, the *Schein*, of what is in itself monstrous.[23]

This is the unity in the *Ineinsbildung*, the imagination as the coming into an image, of the *das Urbild*, the primordial image, and *das Gegenbild*, the counter- or mirror image, the image that reflects the primordial image back to itself without in any way revealing it or making it anything particular in itself. The primordial image presents itself without revealing itself as the counterimage, as if it were seeing without be able to recognize itself in the mirror. In this *Gegenbild* or the counterimage, for example, we intuit the monstrosity of the earth or the ocean without it thereby appearing as a thing and hence just as philosophy "presents the Absolute in *das Urbild*," art presents "the absolute in das *Gegenbild*" (I/5, 369). These primordial images become objective in art and "hence present the intellectual world in the reflected world itself" (I/5, 369). The *reflektierte Welt* is the realm of the *Gegenbilder*, but unlike, say, Lacan's mirror stage, the reflected image does not in any way capture the original. For Lacan, the young morcelated, nonidentitarian body catches a glimpse of itself in the mirror and concludes that it and the reflection are the same. This is the fateful turn: the birth of one's ego, the banishment of oneself to the periphery. One becomes the fixed reference point of the self and thereby also develops "a consciousness of the other that can be satisfied only by Hegelian murder."[24] Schelling's mirror is closer to the mirror in Zen, which takes the perspective not of the one looking in the mirror, but of the mirror itself. The mirror receives and discloses everything without judgment. Each and every *Gegenbild* reflects the great earth, but no *Gegenbild* penetrates or exhausts the earth. The earth is moreover, as John Sallis would have it, an elemental image, monstrously so, its depths shining forth as it meets the sky.

From this perspective, we should be somewhat cautious in ceding the German *Urbild* too quickly to the English *archetype*, at least in the latter's prevailing resonances. In a way, one can say that the elemental depths of the earth are archetypal, but we do not want in saying so to relapse into a Platonism in which the form or primordial paradigm of something is representing itself again and again in its sensible exemplars. Schelling offers some helpful examples, which can also help us appreciate the limits of archetypal thinking. (1) "Music as the primordial image of the rhythm of nature and of the universe itself, which breaks out by way of this art in the world of simulacra" (*der urbildliche Rythmus der Natur und des Universums selbst, der mittelst dieser Kunst in der abgebildeten Welt durchbricht*). Schelling is not saying that Mozart sounds like the breeze rustling through the trees in a forest. Primordial and elemental rhythm emerges in music as *Abbilder*, simulacra, effigies, appearances of what in itself appears as unable to appear, or the echoing forth of the primordial soundless sound. This is not making music by aping the sound patterns of the world. It is much closer to the work of John Cage who could *hear the music in all sound*. In his Tang Dynasty classic, *Shu pu* (*Treatise on Calligraphy*), Sun Qianli regrets that "ordinary people," lacking an "appreciation" that is "deep and subtle," were unable to "recognize the musical potential in the sound of a burning piece of wood."[25] Musical rhythm is not an exemplar of a fixed rule, but the creative coming into being of music as the unfolding primordial song of the earth. (2) The plastic arts, the arts by which, as we have seen in the last chapter, engage in the plasticity and eisemplasy of *natura naturans*, are the "objectively presented primordial images of organic nature itself [*objectiv dargestellt Urbilder der organischen Natur selbst*]." Living plasticity produces the counterimages of the primordial image of the earth. In sculpture, for example, we participate in the great esemplastic splendor of the earth. (3) The Homeric epic as the unfolding of history grounded in the absolute. (4) Finally, "each painting opens or discloses the intellectual world [*jedes Gemälde öffnet die Intellektualwelt*]" (I/5, 369). Painting is not the copying or representation of things that we have seen. It is, rather, a more radical kind of seeing, to see the things of the earth precisely in their disclosivity of the earth. This is the beauty and sublimity of art: "*die Ineinsbildung* of the real and the ideal in so far as it is presented in *das Gegenbild*," which implies: the "indifference of freedom and necessity intuited in something real" (I/5, 383).

One might ask, however, if neither art nor natural science allows us to penetrate the secrets of the earth and lift the veil of Isis, what is their hold on us? What is the draw of what we cannot have, but which nonetheless animates our love for the earth? On the one hand, this question is inseparable from the question as to why we seek to participate with language in the earth's expressivity. Art, for Schelling, is the expressive word of God, "the speaking word of *God*, the logos that is simultaneously God [*das sprechende Wort* Gottes, *der Logos, der zugleich Gott ist*]" (I/5, 483). To love art is not merely to affirm what has been said. It is not the idolatry of the said, but rather devotion to the saying. "But the *real* world is no longer the living word, the speaking of God itself, but rather only the spoken—the coagulated (*geronnen*)—word [*Aber die* reale *Welt ist nicht mehr das lebendige Wort, das Sprechen Gottes selbst, sondern nur das gesprochene—geronnene—Wort*]" (I/5, 484). *Gerinnen* is to coagulate, clot, or curdle (like milk into cheese). The clotted word of art is not the expressive speaking of art. The plastic arts in themselves, too, are therefore only "the dead word [*das gestorbene Wort*]" and the more that this speaking is able to die to itself, the more sublime the artwork is; in music this is perceptible as the moment of sonority (*Klang*), the sounding out of things, the univocity of being, "that vitality that has entered into death [*das in den Tod eingegangene Lebendige*]" (I/5, 484).[26]

Hence, language itself is "vital, expressive, and infinite affirmation," "the highest symbol of chaos" (I/5, 484), the vowels that, as we see in the *Freedom* essay, allow us to hear the silent consonants of nature (I/7, 364). Karl Schelling tells us that his father wrote the following note in the margins of the *Philosophy of Art* lectures: "*Sprache überhaupt = Kunsttrieb des Menschen*: Language overall = the human drive for art" (I/5, 486). That we speak at all is already at the heart of the human drive to creativity, of our desire to participate in the expressivity of nature itself.

Art, speaking at all, is already therefore at the heart of what Schelling understood by that most poorly understood of words, *religion*. In this sense, the philosophy of art is in some way the beginning of what Schelling will later call *positive philosophy*. As Schelling recalled four decades later in 1842 in the tenth and final lecture of his *Historical-critical Introduction to the Philosophy of Mythology*:

> [t]he Philosophy of Mythology also forms an indispensable foundation for the *Philosophy of Art*, just as it has a necessary relation to the Philosophy of History. For it will be essential, even one of its first tasks, for this Philosophy of Art to concern itself with the first objects of artistic and poetic presentations. Here it will be unavoidable to require, as it were, a poesy *originally* preceding all plastic and compositional art, namely, one originally inventing and producing the raw material [*Stoff*]. (II/1, 241)[27]

What was the primordial form of human creativity from which art as such is derivative? A visit to the Lascaux or Chauvet caves, or a study of the petroglyphs that flourished in places as remote from each other as the Americas, Australia, and Tanzania, offer testimony to Schelling's answer: mythology! In the art lectures, mythology "is the necessary condition and the prime matter [*der erste Stoff*] of all art" (§38, I/5, 405). Artistic images emerge from the now forgotten power of mythological images. Artworks and the gods—for "these real, living, and existing ideas are the gods [*diese realen, lebendigen und existierenden Ideen sind die Götter*]" (I/5, 370)—are not tokens for something beside themselves. They do not mean anything other than themselves. *The gods come as themselves.* Schelling insisted that there is *nothing allegorical about the coming of the gods* (see also II/1, 195). Forty years later, borrowing a term from Coleridge, one of the only people that Schelling thought had in any way appreciated the radicality of his single published text on the philosophy of mythology (*On the Deities of Samothrace*), Schelling called mythology *tautegorical.* "To mythology the gods are actually existing essences [*Wesen*], gods that are not something *else*, do not *mean* [*bedeuten*] something else, but rather mean *only* what they are" (II/1, 196; HCI, 136).[28] In fact, the whole *Historical-critical Introduction* takes every possible external meaning to the gods (they were priestly power plays, poetic contrivances, infantile pseudoscience, a magical explanation of what science now appreciates as natural laws, etc.) and renders them all absurd. Nothing explains the gods except the gods themselves.

The gods are the children of the earth's elemental imagination. In the *Philosophy of Art* lectures, Schelling discusses this "exquisite [*trefflich*]" German word *Einbildungskraft*, which he contends actually means *die Kraft der Ineinsbildung*, that is, the force through which something ideal is simultaneously something real (die *Kraft, wodurch ein Ideales zugleich ein Reales*) (I/5, 386).[29] It is the soul animating the body, the force or *principium individuationis* (*Kraft der Individuation*) (I/5, 386). The imagination is "that in which the productions of art are received and *ausgebildet* (educated, formed, cultivated—literally, formed into images) [*das, worin die Produktionen der Kunst empfangen und ausgebildet werden*]" (I/5, 395). To call the gods and artworks imaginative is to understand them schematically, as the "presentation in which the universal means the particular" (I/5, 407).

In this regard we could say that the presentation of the entire endless clamor of the earth is schematic (the revelation of gravity as light, the presentation of the one as the infinite difference of the many). That being said, if the only sway that the gods had over us was that they, like any other being, were schematic, they would not have appeared as gods. The gods are not things among things, but a divine manner of presentation. There is at the heart of the very divinity of the gods something allegorical, but not such that it makes the gods *mean some other thing than themselves*. The gods are also "presentation in which the universal is intuited through the particular" and hence they are simultaneously tautegorical and allegorical, and hence *symbolic*. Their divinity is what makes their forms both what they are and simultaneously elementally expressive. They are not sensuous embodiments of an idea remote to themselves. They are *heautonomous*: *they mean what they say and do not mean anything beside themselves*. To be clear: this is not to mystify the clarity of careful reflection.[30] As Markus Gabriel articulates this: "*Reflection is limited precisely because it is engendered by mythology and not the other way around. . . .* There is no absolute content prior to the mythological form."[31] The miracle of meaning, its *ewiger Anfang*, should be radically and absolutely distinguished from the use of mythology as part of ideology's auto-mystification and its gesture of indirection, that is, its mass distraction from the real. Again Gabriel: "Mythology necessarily arises when we push reflection to its limits. It is only harmful when ideological use is made of it. It can also serve the just ends of radical democracy which does not admit necessary natural conditions at the basis of its laws" (MBR, 78).

Schelling proposed another way of capturing this strange word σύμβολον, originally a token or marker, a ticket, and, in Schelling's repurposed deployment, an elemental simulacrum of the *Ungrund*, a *Gegenbild*, in which the ungrounding ground expresses itself as something, that is, it takes on a pattern or form (the original meanings of image and *Bild*) or look or shape (the original meaning of εἶδος). Schelling turns to another perspicacious and "exquisite [*trefflich*]" German word, namely, the word for symbol: *das Sinnbild*. The word combines image (*Bild*) and sense in both senses of sense. This is not a detached, freestanding "mere image [*das bloße Bild*]," which is "mere *meaningless being* [*das bloße* bedeutungslose Sein]," but rather "the object of the absolute presentation of art should be as concrete, being only equal to itself [*nur sich selbst gleich*], as the image, yet as universal and meaningful [*sinnvoll*, full of sense] as the concept" (I/5, 412). The image is not a picture of itself, but is rather an elemental image, reflecting the depths of the earth.[32] It has therefore an elemental meaning, which, as such, is thoroughly immanent. This elemental meaning, which does not refer to something beyond the image itself, is the image as the presentation of the elemental abyss, yet not such that it is swallowed up and loses its singular face, its concrete, personal *Schein*.

When an image congeals into a fixed meaning, we have idolatry. When I assume that I am what I appear to be—I am *like* I appear—I fail to appreciate the manner in which Schelling understands the force of *likeness*. In his response to Eschenmayer (see appendix), Schelling takes up the question of *das Bild* in relationship to *das Ebenbild*, a precise or spitting image, in the sense that it has in Genesis 1:26 when "God said, 'Let us make mankind in our image, in our likeness.'"

> You scoff that it *falls* to us to make ourselves into the image [*Ebenbild*] of God, to which the understanding also adds its two cents, in that it shows quite artificially how God was actually *forced* to create such a corporeal image of himself. My belief in contrast is that it did not fall to humans to become the image of God, but rather that God himself made the human being in his image, against which it was certainly a different and opposed *Fall* (a Fall of human beings and the devil) by which the human being became the non-image of God. (I/8, 183)

Here Schelling is playing with two senses of *Einfall*. In the first sense it means a "sudden thought," to come or "fall" to thought, that is, for a thought to occur to thinking. We *are* in the image of God, but the loss of this relationship speaks to an ancillary meaning of *Einfall*, namely, the Fall of original sin (in the myth of Eden). One can speak of the *der*

Einfall der Nacht, the fall of night, but here the occurrence is the sudden fall from grace, or what is more typically called *der Sündenfall*. In the *Freedom* essay, this is the fall from the center into the periphery, the flight from our being as *Gegenbilder* of the divine *Urbild*. Schelling, quoting from the beginning of Malebranche's *The Search after Truth*, rejects the claim that the spirit is what informs the body or is in any way on the side of the εἶδος. The spirit is the human *Sinnbild*, so to speak, and Schelling joins Malebranche in his distaste for those who "should regard [*in solchem Irrthum schweben*] the spirit [*Geist*] more as the *form* of the body than as being made in the image and for the image of God" (I/8, 184). Idolatry, the Hebrew *pesel* (פֶּסֶל) or graven image, is the sudden fall from grace and, as such, the loss of one's being as the image of God. The great bifurcation of sense and form that so permeates Western metaphysics has its heart in the Fall and in our inclination toward evil (our striving for form severed from its source).

For Schelling, *das Sinnbild* has sense in both senses of sense. On the one hand, "the meaning is at the same time the being itself [*die Bedeutung ist hier zugleich das Sein selbst*]" (I/5, 411). On the other hand, "the highest charm [*Reiz*] of the gods" is that the gods "*simply are* without any reference [bloß sind *ohne alle Beziehung*]—absolute in themselves—, although at the same time [*zugleich*] they always let the meaning shimmer through [*immer die Bedeutung durchschimmern lassen*]" (I/5, 411).[33] The patency of the earth is able to *durchschimmern*, to shimmer through. It does not come from outside or from elsewhere, but is rather suffused, showing through the gods and artworks like a darker color shows through a lighter color. To *shimmer* is to shine in the manner of a subdued, tremulous light, from an old root also suggesting shadowy, just as we say the first shimmer of moonlight or as Dōgen Zenji spoke of Buddha-nature (*bussho*) as the moon in a dewdrop.

The gods are not interchangeable tokens of something more general. If they are only their ideas (if Zeus, for example, is merely understood by being subsumed into the more general idea that he is a god), they lose the force of their personality (from the Latin persona from Greek πρόσωπον, the masks worn at the theater). Gods are the masks of that which elementally comes only as themselves. Hence, an artwork or a god "should be absolute according to its nature [*seiner Natur nach absolut*]" and "is not there for any end [*Zweck*] that lies external to it" (I/5, 412). Hence, "the poems [*Dichtungen*] of mythology are at the same time meaningful and meaningless [*bedeutend und bedeutungslos*]—meaningful, because it is a universal in the particular, meaningless because both again are absolutely indifferent, such that that in which they become indifferent is absolute and wants to be itself" (§41, I/5, 414). In art, the manner in which the polytheism of the imagination still prevails, we intuit simultaneously the elemental monstrosity of meaning in a world that, as meaningful, only means itself. In mythology we glimpse the "primordial world itself [*die urbildliche Welt selbst*]," "the first universal intuition of the universe" (I/5, 416). From here, echoing the end of the 1800 *System*, we intuit that poesy is the "prime matter [*der Urstoff*], out of which everything emerges, the ocean out of which all streams flow [*der Ozean aus dem alle Ströme ausfließen*]" (I/5, 416).

The Allegorical Reign of the Time of Revelation

However, although the legacy of the gods endures in art as well as philosophy, even though their shadow is even fainter in the latter, the gods themselves, as Hölderlin also acknowledged, have died. Christ is the last god, but also the one who comes to found a new world and to annihilate (*vernichten*) and "absolutely close" the world of the gods (I/5, 432). The ascendance of revelation evacuates the world of the gods because it eradicates the reality of divine mythological objects. Christianity in turn is an idealistic mythology, retracting the finite back into the infinite and demoting the former to a mere *allegory* of the latter. It allegorically suspends (*aufhebt*) the symbol, leaving in its wake only symbolic acts (e.g., the sacraments), but not symbolic objects. The exception was mysticism, the intuition of "the infinite in the finite" (I/5, 447), but the mystics were generally regarded as apostates and heretics, and, in the end, mysticism is too personal, too subjective; it is "merely inward" (I/5, 455), and hence produces only "subjective symbolism," which as such is "in and for itself unpoetic" (§56, I/7, 456).

In the later Munich and Berlin lectures, Schelling would sharply delimit the range and force of the mystical, lamenting its incapacity to know or even to speak. Speaking, for instance, of the mystical strain within theosophy, Schelling chided: "If they were really in the Center, then they would have to go silent, but—they want to talk at the same time, to speak out, and to speak out for *those people* who are outside the Center. Herein lies the contradiction in theosophy" (HMP, 187/181).

The mythological is the poetic expression of the silent heart within revelation. It is not enough merely to intuit it—this is merely the limiting case of all negative philosophy. Even speculative physics is in the end a kind of *objective mysticism*, for it, too, "intuits the infinite within the finite but in a universally valid and scientifically objective manner" (I/5, 448). The historical dimension of what Schelling later calls positive philosophy attends to the catastrophic succession of religions (gods, like all beings, go the way of dinosaurs). Even positive philosophy, with its careful genealogy of the past and its radical discernment of the present, cannot forgo the "holy of holies" of the creativity of inspired, living art. Art, as part of the new mythology that affirms all things human and the great elemental prodigality of the earth, is the coming into the finite of the infinite that counters revelation's lopsided alignment with the allegorical. In philosophical religion, one could surmise that the new mythology would liberate not only the devotional patience of science, but also the unprethinkable life of art works. The genealogical investigation of the past, beginning quite dramatically with *On the Deities of Samothrace*, is not only narratival but also utopian as it unleashes the promise of a new future sublimated within the fossilized religiosity of the past.

Beginning with the *Philosophy of Art* lectures, the strength of Christianity in particular and of revelation more generally, is its allegorical and therefore "unconditional devotion to the immeasurable [*die unbedingte Hingabe an das Unermeßliche*]" (I/5, 430). As the esoteric, ungrounding ground of existence, it reveals its *historical* dimension (each being is an expression of the divine while at the same time revealing nothing about the divine in itself). As the symbolic realm shrinks to mere symbolic acts, the power and force of the exoteric dissipates, reducing the earth to a gateway beyond itself. For example, although the Eucharist is an allegory of God, in the end also thereby devalues the sacredness of bread in its own right. The point of the Eucharist is that bread can be transubstantiated into a presence that has nothing to do with bread. Wine is no longer the liberating presence of Dionysus, but rather the presence of something remote to itself. The intuition of the universe as the kingdom of God makes it an expression of an absolutely removed absolute, and hence a devaluation, abdication, and even annihilation of this world for the sake of the otherworldliness of the ideal. Catholicism lacked a compelling mythology and although Luther intervened in this world, he, too, despite his insistence on the emptiness and groundlessness of the very form of Christianity, was unable to transform it into a new mythology with new, genuinely external, forms to make it objective. When the Good is only otherwise than being, than the products of the earth are not in themselves worthy of cherishment.

One might even here note how much Schelling's delimitation of Christianity as lopsidedly ideal resonates with Nietzsche's own critique of it, despite the latter's utter disavowal of the possibility that anything transvaluative remained within Christianity. A thinker no less than Georges Bataille discerned that Nietzsche was much closer to Jesus than the former was prepared to admit. Speaking of the "deep kinship between Nietzsche and Jesus," Bataille reflected that, "above all, both were moved by the feeling of sovereignty that possessed them and by an equal certainty that nothing sovereign could come from *things*. Nietzsche sized up this resemblance. He was in a position opposite Jesus!"[34] For both Schelling and Nietzsche, only a god can save us, but by that we mean: our salvation does not lie in a detached transcendent realm, but rather with liberating the depths of creativity. "The prerequisite of a mythology is precisely *not* that its symbols mean [or refer to] mere ideas, but rather that they are independent beings that refer to themselves" (I/5, 447).[35] This is why a new mythology must be created (*erschaffen*)—allowed to come into being—and cannot just be derived from the instructions (*Anleitung*) of philosophical ideas. If it is not created, it cannot be given "an independent poetic life [*ein unabhängiges poetisches Leben*]" (I/5, 446).

Schelling provided some hints of works that spoke from the creative sensibility (in both senses of sense) of a new mythology of revelation. There was Dante who composed his "divine poem" out of an ignorant, intellectually half-baked world, "the barbarism and even more barbaric learning of his time" (I/5, 445). Ugolino the mythological symbol is now more real than his historical precedent. Or Shakespeare's Hamlet or Lear! Or Cervantes' Don Quixote and Sancho Panza—who in the early seventeenth century is more *vitally real*? Or one could also look to speculative physics, which can also be poetic if handled imaginatively and with originality, which is not to be confused with *mere particularity* (affected and willful eccentricity, posing peculiarity, doing strange things just to be strange, etc.) "Each originarily handled matter is precisely thereby also universally poetic" (I/5, 447).[36]

The distinction between mythology and revelation, as Schelling will later develop it, is a distinction between *necessity* (gods coming as themselves) and *grace* (free revelation). We could already say that without history, that is, without the intuition that gods and works, like all beings of the earth, are posited temporarily precisely insofar as positing as such is temporality (I/2, 364), gods and works again become mere idols. When this happens, we forget that we are made in the image and likeness of God, not in the image and likeness of anything in particular. In history "they first become gods [*Götter*], but before they were idols [*Götzen*]" (I/5, 448) and only as gods do they become "truly independent and poetic" (I/5, 448). Only in history—the temporality of matter—are they monstrous images of the divine, images in a dynamic unfolding cosmic poem.

The task is not to graft the "real" Greek gods onto the allegorical otherworldliness of Christianity. Not only were the gods really real, they are also now really dead. The task of a new mythology is to again sow the seeds of their *ideality* into nature. Neither philosophy nor physics can produce the coming gods. We can only "await its gods," knowing that "we already have the symbols in readiness" (I/5, 449). We can only await the coming of this *Zumal*, this simultaneity or "at once" where freedom comes as necessity and sovereignty comes as form, bringing together nature and history (I/5, 449) so that the gods, having become the gods of history, become more fully the gods of nature (I/5, 457) and the images of the earth.

The New Mythology of the Whale

"But it is this *Being* of the matter; there lies the knot with which we choke ourselves. As soon as you say *Me, a God, a Nature*, so soon you jump off from your stool and hang from the beam. Yes, that word is the hangman." This passage concerning the separation of existing things from their ground, from Melville's famous Spring 1851 letter to Hawthorne,[37] resonates profoundly with Schelling's thought. As soon as being is only the problem of the absolute *Ich*, as Schelling charged it was with Fichte, or identifies with God as the supreme Being as it does in what Heidegger was later to call onto-theology, or reduces to a crude scientific positivism, which was increasingly the fate of science in Schelling's century, it is *game over*. Moreover, Melville, in confiding with

Hawthorne about his strange new work—*Moby-Dick, or the Whale*—in November of that year, confided that "I have written a wicked book, and feel spotless as the lamb. Ineffable socialites are in me." It appears like the work of the devil, but it is as immaculate as if Christ himself had written it. As we have seen, Schelling was convinced that, with the exception of Dante's *Divine Comedy*, there has not yet been a genuinely Christian art that has the symbolic force of the mythological. (The aborted project of *The Ages of the World* was in part an experiment in a new philosophical-poetic *Divine Comedy* for modernity.) Too much contemporary art in the monotheistic world lacks the life of pagan art and appears merely contingent, prompting us to ask the easy question, "Why, for what reason is it there?" (II/1, 242; HCI 168). The arbitrary multiplication of images does not get at the problem of the life of images, "namely, the necessity of actual beings [*wirkliche Wesen*], which do not merely signify [*bedeuten*], but rather *are*, and which at the same time are *principles, universal and eternal concepts*" (II/1, 242; HCI, 168).

The coming of the whale must be the coming of a living whale. As Charles Olson framed it: "the job was a giant's, to make a new god."[38] Ishmael, steeps himself in cetological minutia because Moby Dick is not a token of a remote idea, a sensuous and concrete representation or illustration or exemplification of a remote idea about what a whale really is. *Moby Dick comes as himself*, emerging not as some general spermaceti whale, but rather as having a proper name and the singularity that such naming suggests. "So ignorant are most landsmen of some of the plainest and most palpable wonders of the world, that without some hints touching the plain facts, historical and otherwise, of the fishery, they might scout at Moby Dick as a monstrous fable, or still worse and more detestable, a hideous and intolerable allegory" (chapter 45, The Affidavit).

Although Melville did not read German and hence there is no reason to conclude that he had read the *Freedom* essay, he did esteem Coleridge, including the latter's explicit embrace of Schelling in his *Biographia Literaria*, and the indirect influence is patent. As Sanford Marovitz remarks, the resonances and affinities of Schelling via Coleridge are "little short of astonishing" and they "document the dynamic effects of the *Zeitgeist* upon the fertile literary mind":[39]

> The correspondence between Schelling's *Of Human Freedom* (1809) and *Moby-Dick* provide a case in point. Not only

> elements of the thought but at times even the images are similar. In that tract, Schelling explains in very complex and paradoxical language in which cosmic order and reason have been created from chaos and unreason; although disorder was then submerged, a residue remains below, Schelling depicts this with images of light and darkness—what would be for Melville "love and fright"—showing how man's conceit (which Melville might have personified in Ahab) is disruptive.[40]

Although I think that it is not quite right to speak of Schelling's creation as if it were a kind of agency (the classical creator God who acts from a realm independent of being) nor to speak of images as if they were somehow pictures of ideas—this is not the case in either Schelling or Melville—, Marovitz seems to me nonetheless to be spot on to detect that *Moby-Dick* can be read as the great American Schellingian novel and that Ahab recasts the tragedy of evil in a specifically New World spacing. The latter becomes all the more striking given that the novel largely transpires in the wide-open spaces of the sea, not within the confined norms of the land and its human cultures: "Pip came to the surface mad, Melville possessed of his imagination. The Pacific gave him the right of primogeniture" (CMI, 116). Indeed, Melville, Olson also tells us, "had a pull to the origin of things, the first day, the first man, the unknown sea. . . . He sought prime" and attempted to "make a myth, Moby-Dick, for a people of Ishmaels" (CMI, 15). It is outlandishly American, extra-normally American, or, to use a phrase that Deleuze himself associates with Melville, it is *deterritorializing*. "Over Descartian vortices you hover. And perhaps, at mid-day, in the fairest weather, with one half-throttled shriek you drop through that transparent air into the summer sea, no more to rise for ever. Heed it well, ye Pantheists!" (chapter 35, The Mast-Head).

Melville's great open sea, which swallows everything without a trace and out of which emerges ceaseless wonders, including the monstrous Leviathan, is an *Urbild* of the χώρα. "No mercy, no power buts its own controls it. Panting and snorting like a mad battle steed that has lost its rider, the masterless ocean overruns the globe" (Chapter 58, Brit). The sea is sovereign ("masterless"), the ungrouding that suspends the grip of terrestrial habits of being and acting. As Job, an obvious influence on Melville, realized shortly before being smitten with boils, Yahweh gives and Yahweh takes away (Job 1:21). This, too, was the primacy of the sea and the primacy of the Leviathan (who Yahweh evokes when speaking as the whirlwind). This for Melville is the thought of the eternal beginning: "Melville wanted a god. Space was the First, before time, earth, man. . . . When he knew peace it was with a god of Prime" (CMI, 82).

Ahab, like Schelling's fallen human who inverts the relationship between good and evil by making themselves the ground, aspired to make the power of the sea his own. Ishmael characterizes Ahab as beset with an "unexampled, intelligent malignity" and an "intangible malignity" and "monomania." He exhibited a "special lunacy," which "stormed his general sanity, and carried it, and turned all of its concentrated cannon upon its own mad mark" (chapter 41, Moby Dick). Ahab, overcome by *ressentiment* for his own pettiness, can only see the great Whale. "He piled upon the whale's white hump the sum of all the general rage and hate felt by his whole race from Adam down" (chapter 41, Moby Dick).

Although the *Freedom* essay remains the last published analysis of the problem of evil in Schelling's lifetime, he did not altogether abandon the theme and his return to it decades later in the *Philosophy of Revelation* sheds additional light on the Ahab within us all. This time around Schelling concentrates on the mythological image of Satan. There is no such thing as a devil, no bizarre Manichean creature dedicated to leading human beings astray. *Satan is not a being and has no being of his own*, but that is precisely the problem. He is the temptation to-be-for-oneself as such. He is the Prince of Lies insofar as he tempts humans to break from the center and to establish themselves as their own center, to become the "lord of human consciousness so that the human becomes his slave" (U, 644).

This is Satan's lie: that he should have being, that the center should be able to take ownership of itself. Satan, "that which does not have being, but which, through deception, desires to be able to be a being" (U, 644), is the errancy of the ungrounding ground seeking to get a hold of itself and become a being, to represent itself to itself in the mirror. This is the doubling at the heart of the primordial image of Satan: "an infinite source of possibilities [*Möglichkeiten*] in himself" and "his incapacity [*Unvermöglichkeit*] to actualize these possibilities, out of which emerges his infinite craving [*Sucht*] for actuality. This spirit was imagined [*vorgestellt*] as conceivable as an eternal neediness [*Bedürfnis*] to actualize through the human will what is in Satan a mere possibility" (U, 646). *Satan is the potency—the hunger to be—that tempts human beings to imagine that they are gods.* The hunger to be is now a being, and that being is me! *I am the center.* Although Satan is the "sophist par excellence," without this "eternal hunger and thirst" (U, 644), there would be no history, just eternal sleep. Goethe's God in *Faust* accepted and tolerated Satan even though Goethe's Satan was the possibility that Faust himself would in his quest for knowledge usurp the center. God in the *Book of Job* also understood that Satan was a necessary part of Himself, the part that roves and roams (Job 2: 1). In the same way, Schelling's God (following John) shares the universe with sickness—"sickness only came into the world through Satan" (U, 652)—and evil. Only in the latter, only in having lost the ungrounding ground, can God be revealed to us. We cannot begin at the beginning, but only find health and the sacred from sickness and hubris.

Although Ahab would be Leviathan, it comes at an exorbitant cost, namely, the loss of all else that he is, and, in the destructive solipsism of obsession, the sacrifice of the Pequod. Ahab, humiliated by the sea—by the watery unruliness of the χώρα, so to speak—sees in its unfathomable expanse his humiliation. Ahab in himself is nothing and what Schelling in the *Freedom* essay calls *die Angst des Lebens* (I/7, 383), anxiety before the monstrosity of life itself, drives him to the periphery where he seeks to subjugate the sea and the great leviathan to himself.

Hence when Ahab spoke to the silent severed whale head as if it were the sphinx, demanding that it speak in return, he received only monstrous silence. "O head! thou hast seen enough to split the planets and make an infidel of Abraham, and not one syllable is thine!" (chapter 70, The Sphinx). Indeed, "for unless you own the whale, you are but a provincial and sentimentalist in Truth . . . What feel the weakling youth lifting the dread goddess's veil at Sais" (chapter 76, The Battering-Ram)? Ishmael is here alluding to what Kant in the *Critique of Judgment* deemed language's most sublime accomplishment: "Perhaps there has never been something more sublime said or a thought expressed more sublimely than that inscription over the Temple of Isis (of Mother Nature): 'I am everything that there is, that there was, and that there will be, and no mortal has lifted my veil.'"[41] Ahab is the one who becomes exclusively defined by his fixation with lifting the veil and harnessing its power for himself, seeking to dominate what the whiteness of Moby Dick evoked: "all deified nature absolutely paints like the harlot, whose allurements cover nothing but the charnel house within" (chapter 42, The Whiteness of the Whale).

Staring at his Ecuadorian coin, Ahab muses, "There's something ever egotistical in mountain-tops and towers, and all other grand and lofty things; look here,—three peaks as proud as Lucifer. The firm tower, that is Ahab; the volcano, that is Ahab; the courageous, the undaunted, and victorious fowl, that, too, is Ahab; all are Ahab" (chapter 99, The Doubloon). All that towers above the valley shall be made subject to the valley. That is how the vengeful valley seeks to live with mountains. This is confirmed as Ahab, against the lightning-filled sky, screamed, "I own thy speechless, placeless power . . . I am darkness leaping out of light, leaping out of thee" (chapter 119, The Candles). When Ahab becomes the power of the sea, there is only "OVER ALL, hate—huge and fixed upon the imperceptible," a "solipsism which brings down a world" (CMI, 73). And what is this hate, this Satanic force where there is only oneself, not the solitude of God (chapter 2), but the lonely world destroying solipsism of the ego? "Declare yourself the rival of earth, air, fire, and water!" (CMI, 85).

In direct contrast with Ahab's extra-human monomania is Pip's madness, which is as wide as the vast expanses of the sea because it no longer seeks to subjugate it to his ego. "He saw God's foot upon the treadle of the loom, and spoke it; and therefore his shipmates called him mad. So man's insanity is heaven's sense; and wandering from all mortal reason, man comes at last to that celestial thought, which, to reason, is absurd and frantic" (chapter 93, The Castaway). Indeed Pip, who caws like a bird, has died to himself, and later tells the shipmen that he "died a coward" (chapter 110, Queequeg in His Coffin). Ahab at first becomes fiercely protective of Pip, sensing in him not only their shared madness, but also the vulnerability out of which madness issues. Eventually, however, Ahab must reject Pip because openness to space no longer mediated by the ego is radically incompatible with Ahab's drive to be Ahab. "Weep so, and I will murder thee! have a care, for Ahab too is mad. . . . And now I quit thee" (chapter 129, The Cabin). It was all just too much, as Ahab's poignant, lonely final words to Starbuck—for what can be lonelier than a loneliness that drives one to the hermetic loneliness of solipsism?

> Forty years of continual whaling, forty years of privation, and peril, and storm time! forty years on the pitiless sea! for forty years has Ahab forsaken the peaceful land . . . —o weariness! heaviness! Guinea-coast slavery of solitary command! . . . away,

> whole oceans away, from that young girl-wife I wedded past fifty, and sailed for Cape Horn the next day, leaving but one dent in my marriage pillow—wife?—wife?—rather a widow with her husband alive! . . . mockery! mockery! bitter, biting mockery of grey hairs. . . . (chapter 132, The Symphony)

And so Ahab becomes trapped in the hell of being Ahab—"Ahab is for ever Ahab, man! This whole act's immutably decreed" (chapter 134, The Chase—Second Day). In dying he confesses his desolation: "Oh, lonely death on lonely life" (chapter 135, The Chase—Third Day).

Moby Dick swims away, unvanquished. Before the life of imagination, we see, as we developed at the end of the fourth chapter, Schelling's three great human possibilities. Ahab is the tragedy of the great sacrilegious *Verstandesmensch* who would subjugate all of the earth exclusively to its comprehensibility. There is no price that is too high when one is poisoned and sickened by such mania, and the dire state of our earth that we discussed in the opening chapter of this book bears this out. The second possibility is Pip, whose mind becomes supremely vulnerable as it is disintegrated when it becomes as vast as the sea. This is the way of the mad and of gods. Beyond the monomania of Ahab and the χώρα-mania of Pip—the madness of the minute and the sublime—lies only Ishmael. Melville and Schelling, each in their own way, wrote for a new earth of Ishmaels. Here is the renewal of art, philosophy, and religion as each in their own way negotiates the madness and gravitational monstrosity at the heart of the imagination. "For in what does the intellect prove itself than in the coping with and governance and regulation of madness?" (I/8, 338).

APPENDIX A

Schelling's Answer to Eschenmayer (1812)

TRANSLATED BY
CHRISTOPHER LAUER AND JASON M. WIRTH

[161] Answer to the Preceding Correspondence[1]

Munich, April 1812

If we were not separated by space, your letter and my answer could perhaps have made a dialogue. Despite this distance, I would like as much as possible to give our encounter such a form and believe this easiest to achieve if I use marginal numbers to indicate the places in your letter that seem to me to demand an answer before the others. By these means the reader will find himself in position to hold the two discourses up against one another.

I have often wished that, as in ancient times, today, too, public dialogues could take place in the presence of learned witnesses, if not over articles of faith, at least concerning philosophical contentions and systems. How much would be gained thereby, and not merely in connection to those [discussions] that aim only to obscure and confuse the concept so that no one can see clearly any longer or learn from their perspective. But also between genuine inquirers into the truth of things the beginning of a contradiction arises not infrequently in wholly unobserved discrepancies, often in an innocent-appearing alteration or substitution of concepts that eventually leads to the most monstrous of consequences, at which the honest opponent must himself be astonished. In verbal interaction such confusions are confronted right at the beginning and their substance destroyed just as they emerge, thus warding off the long polemical [162] discourses and arguments that so often unfold despite being at base worthless tirades.

You have agreed to have your correspondence printed together with mine and have thus let it be known that your goal is determining the truth, or at least the actual point of contention, which cannot be easily concealed in such an immediate juxtaposition of objection and answer.

Now as *you* have determined to present your exceptions and comments directly and in an assertive tone, you will find the same in my language (where it is appropriate to the matter at hand), without encountering in the openness of my expression any injury to your long honored respect and friendship.

Incidentally, do not expect me to follow the order of your letter exactly. For it is easy to accumulate objections without following a methodical path, but the answer, since it can only follow from a view of the whole, must already be systematic in a certain sense.

I distinguish your counterarguments against me from the claims you make in your own voice. I will first attempt to answer the former and then will share with you my thoughts on your manner of thinking.

First, you are not at all content that I regard scientific investigations of the essence [*Wesen*][2] of freedom as possible and have actually undertaken them. To you the essence of freedom, like its relationality, seemed always already caught up in the sort of demonstration you take to define science in general. Freedom, you say, could never become a concept [*Begriff*]—as if anything could become a concept that is not one already! It seems to me that you could say the same thing in the same sense of a stone. Likewise, when you ask if the ethical, virtue, and beauty are *thought*, I could ask in response, is the stone, is tone, is color *thought*? In this respect freedom is nothing unusual.

[163] It almost seems that you believe that what is made an object of scientific investigation is by that very fact already made into a mere concept, as you say (15) that since I call freedom a concept, it must be mixed up for me with necessity. But I do not know that I call freedom a concept, though I do speak of a concept of freedom. But I can also

speak of the concept of anything at all, e.g. a stone, without thereby passing off as a concept the stone with which one builds houses (16).

Your argument on this matter would thus have the flaw that it proves too much. You acknowledge that one must acquaint oneself with freedom and thus make a concept of it. But that would be precisely the irrational [*das Irrationale*] *of which I speak.* Freedom can never be taken up fully into the concept, and there must always be a remainder that does not resolve into the concept [*ein Rest, der im Begriff nicht aufstehe*].[3]

You want to seek the irrational in the heights and I want to seek it in the depths. You call irrational what is immediately present to our spirit, like freedom, virtue, love, and friendship (17), even though you are not perfectly consistent in this use of language. In other passages (6) you apply this concept in a sense completely analogous to mine. I call irrational what is most contrary to the spirit, being as such [*das Sein, als solches*], or what Plato calls "what does not *have being* [*das Nicht*-Seiende]."[4]

Should this contradiction not already be decided merely by an exact determination of the concept? The expression "to dissolve [*auflösen*] into concepts" has become familiar with frequent use; thus it could easily happen that it is applied in a perverted sense. The *concept* of a matter [*Sache*] is indeed nothing but the spiritual becoming-conscious [*das geistige Bewußtwerden*] of it. *Thinking* is nothing other than the spiritual process by which we achieve this becoming-conscious. Dissolution in the chemical sense (from which the expression is taken) only takes place where something of a different constitution comes to be of the same constitution as its solvent. But for us to become spiritually conscious of freedom, virtue, friendship, and love [164] requires no special operation of incorporation or dissolution into consciousness; they are in themselves of the same sort as the spiritual; concept and thing [are] immediately one. On the other hand, what resists the spiritual or thinking, i.e., the real, being as such, this "is" of which we can indeed be spiritually conscious—its concept consists, however, in *not* resolving evenly [*aufgehen*] into the concept. It is thus in my reckoning a complete reversal of the true relation to call the spiritual irrational and the non-spiritual rational.

Now that I have dispatched these general arguments against me, please allow me to move on to the objections you level against my essay's basic concepts of *God's nature and essence*.[5]

"In God (that I speak of the existing God is self-evident) one must distinguish between the mere *Ground* of existence [*Existenz*] and existence itself or the *subject* of existence." So goes my claim. To counter it, you state (1) "Because God has the ground of his existence in himself, the ground immediately ceases to be ground and *collapses into one with existence*."—This I grant completely. The ground of existence and existence are in themselves not distinct if you understand by the latter nothing more than existence, pure existing as such. But if by "existence" you conceive the whole, insofar as the existing subject belongs to it, then I must deny it, for the ground is the non-subject, not-being-itself [*das nicht selber Seyende*], and is thus distinct from existence insofar as the subject is conceived together with it. I have not spoken at all of a difference between *existence* and the ground of existence, but of a difference between *what exists* [*das* Existirende] and the ground of existence, which, as you see, is a significant difference.

Here you will probably say the following: Because God has the ground of his existence in himself, this ground must *collapse into one* with the existing God, or else that (2) in [165] my doctrine essence and form and similar oppositions collapse into a single point.

"Collapsing-into-one" or "-into-a-single-point" are expressions that have unfortunately been made somewhat airy through frequent use. When you say "collapsing into one," do you mean *becoming one and the same* [*einerlei*], or *belonging to a single essence*? You seem to want to conflate the two concepts. In the first sense your inference would be incorrect, and in the other I would agree with you. God has the ground of his existence *in Himself*, in his own primordial essence [*Urwesen*], to which the existing God (God as subject of existence) also belongs. I describe this original being, from which God himself first emerges in the act of his manifestation, clearly enough in my essay (I/7, 406), and in order to distinguish it from *God* (as mere subject of existence), I have not called it God, but the *Absolute* pure and simple. But from the fact that the ground and the subject of existence belong to a primordial essence, it does not follow that they are not at all distinct. Rather, precisely because they belong to a single essence, they must be distinct, even opposed in another respect. A human being's mind [*Gemüth*] and spirit belong to one essence and actually collapse into one in regard to the determinate person; that is, together they just constitute the single primordial essence, from which the actual person is only the unfolding. But in this unfolding there are necessarily distinct and mutually independent potencies that can, as is generally known, come into contradiction with one another.

But not merely the concept of *ground* (which, by the way, you take not in the way I have thought it, but in the vulgar sense, in that you (2) attribute to it as a correlate the concept of *consequence*, according to which finally the God which has being [*der seyende Gott*] would be a *consequence* of being [*Seyn*], *of* the ground *which does not have being* [***des nicht-seyenden*** *Grundes*])[6]—not just the concept of ground but also the concepts of form and essence, being and becoming, all concepts of the understanding overall, are for you, when applied to God, inappropriate and condemnable (3), in that God is thereby made into an object of the understanding [*Verstandeswesen*], [166] or as you express it later (4), into a merely particular God.

With this our work has come rather far afield and in a certain way extended to the general question of the value and the validity of concepts of the understanding in relation to God. On this disputed question I cannot get involved, first and foremost because I am only addressing what concerns my arguments in particular and do not wish to undertake correcting your concepts regarding subjects on which you are in contradiction with *generally* accepted ideas [*Vorstellungen*]. I believe that it is easy to show that this is the case, and you yourself openly admit as much (14). Is not God necessarily already in a certain sense a particular being [*Wesen*] precisely because he is a personal being? Can personality be thought without isolation, without being-for-itself, thus in this sense without particularity? Further, can a being acting according to purpose and intention be thought that is not *eo ipso* also an object of the understanding?

I thus praise you openly for the uncommon consequence and openness with which you reject the one like the other and say directly that we are not *justified* in *applying* the predicates of omnipotence, benevolence, and omniscience to God (10). And the same is true with the predicates of self-sufficiency, personality, self-cognition, self-consciousness, life, etc. At the end [of your letter] you even forbid saying of God that He is, suggesting that all of these are only anthropomorphisms [*Menschlichkeiten*], particularities created by our minds and processes of thought, inappropriate to God's dignity. According to these claims, God does no better than the Oriental monarchs, who are robbed of all free movement and human expression under the pretenses of their station above everything human and the honor their citizens give them as divine. In order rightly to elevate God and place Him at a distance from everything human, you carefully remove from Him all sensible and comprehensible properties, forces, and effects.

[167] If followed consistently, all Kantian, Fichtean, and Jacobian philosophy, that is, the whole philosophy of subjectivity that rules our time, must come to your conclusion. In this respect you alone have escaped the self-deception that deludes others. True, for different reasons they possess a great, if not always understood interest in the doctrine of a personal God. But in order to conceive this doctrine, presuppositions are required which offend the purism to which our world-wisdom aspires. Only, as *you* have rightly felt, here there is no choice. There is either no anthropomorphism whatsoever and therefore no representation [*Vorstellung*] of a personal God that consciously and intentionally acts (which would have made him entirely human) or an unlimited anthropomorphism, a thoroughgoing and *total* humanization of God (with the single exception of God's necessary being). This is what makes some who would like to be seen as professional philosophers balk. For it is familiar to every beginner in the rational study of religion that God is elevated high over everything human, and Kant has indisputably proved that any application of concepts of the human understanding to God is impermissible and misguided. They thus believe they could bring themselves the renown of philosophers if they explained how a personal God could actually be conceived. But they only want to talk about such things because it sounds nice and gives off an air of edification.

Moreover, as I said before, I do not want to get involved with this disputed question. My main reason is because I grant no validity at all to this entire type of argumentation. The primary question cannot be with what right we apply our concepts to God; *before all of that we must first know* what God is. For granting that continuing investigation finds that God is actually self-conscious, alive, personal, in a single word, humanlike, would there still be an objection that we apply our human concepts to him? Now if he is in a way human [*menschlich*], who can object? If, as you say, my reason [168] has placed itself above God in what you affirm of God, then your reason would do the same that much more decisively in what you deny of God, in that [your reason] allows itself a priori, without any investigation, merely subjectively, to pass judgment on God. For I do the opposite and claim nothing out of myself, but rather seek to follow after *His* ways.

Then on which side does the arrogance (under the semblance of humility) lay? Which of the two hides the conceit of human judgment under the shell of meek denials?

You say that God *must* in all cases be superhuman. Now if He wanted to be human, to return to you a question that you ask on another occasion (12), who could object? If He Himself climbed down from on high and *made* Himself *common* with His creation, why should I hold Him on high through force? How should *I* debase Him through the representation of His humanity, if He Himself were to debase Himself?

Before objective investigation discovers what God is by development of the primordial *Wesen* itself, we can just as little deny something of God as we can affirm it. Indeed, what he is, is through *Him Himself*, not through us. Thus I also cannot dictate in advance what he should be. He is what He wants to be. Thus I must first seek to ascertain His will, but not to resist in advance that He is what he wants to be.

This entire line of argumentation is as out-of-date as the Kantian philosophy and should rightly no longer be heard. If we say that God may not be thought with human concepts, then we likewise make creation into one of our human concepts, only as a negative measure of the Godhead [*Gottheit*], as Protagoras made what is actually merely subjective in human beings into the measure of all things. But on this you go even farther than Kant and make the *earth* (5), or, as you call it, our "miserable earth-spheroid"—it does not seem to me so utterly contemptible—again the measure of our concepts. [169] None of our spiritual thoughts transcends the earth.[7] The application of our root concepts would be invalid on the nearest planets or the sun (6), to say nothing of heaven. Thus if God is self-conscious, self-cognizing, and personal, then this is for Him (11) a being-affected by the concepts of our earthly understanding (which He nevertheless has planted in us), and in order to prevent this being-affected, He must renounce life and personality.

Moreover, for the reasons presented, I cannot take up the argument that you take from your theory of reason and understanding, in which the latter appears sometimes as the prism for the other-worldly, indifferent ray of light (8) and sometimes as evidence of a doubled reflection (9), for I find it utterly reprehensible to wish to speak [*absprechen*] about God for reasons taken not from Him Himself, but from our faculties of cognition. But furthermore, I can only view those propositions about reason and understanding as entirely arbitrary assurances. I therefore move from these general arguments, which all presuppose what must be proved in the first instance—and thus, to tell the truth, prove nothing—to the remarks in your document that are directed against the *evolution of God from out of Himself*, as they are in my essay not so much presented as alluded to.

"I give God," you say, "the longing [*Sehnsucht*] to give birth to Himself, as if there could be in God a wish to be something that He is not already." To this you then argue (13) that longing is something completely human.[8] You could have spared yourself this argument in every respect, however, if it pleased you to reproduce my thought with the terms given in my essay instead of translating it with the nearest arbitrary expression. I maintain there is in God, but outside of that which *He Himself* actually *is*, a different principle distinct from this [selfhood] but not separable from it, whose character is longing, and indeed the determinate longing to receive the divine in Himself and [170] present it. From precisely this it is shown that it is the sole instrument of revelation and actualization (setting-into-activity) of the proper subject or being [*Seyenden*]. Longing is his *character*, like that of all of nature, not in that nature might have a longing, but in that it is itself and essentially longing. If it were possible (which indeed is not the case) for you to be permitted first to confuse or *conflate*[9] this being of God's proper self with God Himself (as being [*seyenden*]), then you yourself could not say that I ascribe to God a longing, or I grant Him the same, as one could say of a *property*. You must at least then say that I make God Himself into longing, or I claim that God Himself is essentially longing. Thus, however it is taken, the expression you have used is imprecise and inappropriate to my representation. And the nerve of the argument, that this principle is the single implement of the actualization of the concealed essence [*Wesen*], which in itself has the being of itself of the Godhead, is not touched at all.

Still, you ask, to what end is this longing that indeed directs itself at the understanding, yearns for it, but is not itself understanding and is therefore without understanding (I/7, 360)?[10] "If God was from eternity, how should the lack of understanding precede understanding, chaos order, and darkness light?" (35). Indeed, I wonder how you can express the "*was*"[11] of God after you had just before forbade Him the "*is*" (27). I thus content myself with asking you in what sense you take this "*was*." If God *was* from eternity—i.e., in the state of revelation and of completely external actuality, in which He posits himself only through the completion of creation—then it is admittedly not apparent how chaos ever could have preceded order, the lack of understanding could have preceded understanding, or darkness could have preceded light. But if God was not from eternity—i.e., in the state of actuality and revealed existence, if He in contrast performed a *beginning* of his revelation—then this does not yet prove that in the present case order was preceded by chaos, light by darkness, [171] understanding by lack of understanding. But no objection arises against this that would be taken from God's eternity. You now have the choice of which you would rather admit: that God was from eternity, namely in his original essence and not-revealed being [*Seyn*], or that He was from eternity revealed being [*Seyn*]. In the first case your argument has lost its nerve; in the other you claim that creation is equally eternal with God or that God has necessarily created from the beginning. But then your objection does not especially or uniquely address my argument. And in the same sense you could oppose the general doctrine, which Christianity also holds, that there is a *beginning* of divine revelation, by saying, "The eternal could not tolerate time or in itself bear a change in things."

And yet you attack that explanation of creation from a more dangerous side, the moralistic one. Because I maintain that what is without understanding is *historically* or actually before understanding and darkness before light, do you intend (34) (which could hinder us) to push this explanation further and allow virtue to emerge from vice, the saintly from sin, heaven from hell, and God from the devil? For you add: "What you call the dark ground of God's existence is actually something similar to the devil."[12]

I wish you had not made that last remark. I would have hoped you would not read my essay in such a facile manner if you wished to pass judgment on it. With only a modicum of attention, passages like the following could not have escaped you: "We have explained once and for all that evil as such can *only* arise *in the creature*" (I/7, 374); "The original grounding essence (under which precisely that dark ground of existence is understood) can never be evil in itself" (1/7, 375); "It can also not be said that evil comes from the ground or that the will of the ground is the author of evil" (I/7, 399). Such a false attribution, whose [172] context makes it less an objection than an actual argument that according to my basic principles God emerges from the devil, should be left by a man of your spirit to those whose weak understandings can be led against their will by strange and unfamiliar concepts to such *monstra et portenta* [monsters and wonders].[13]

I thus adhere simply to the presupposition contained generally in your words that God emerges according to that theory from a ground that is either the same as Him or at least comparable.

You are aware, and even point toward a corresponding passage in my essay, that I posit the primordial essence [*Urwesen*] or the non-beginning and absolutely eternal Godhead *before* and *above* all ground (I/7, 406) and in this connection always speak of the ground as a principle that God has in himself (in his original essence). Thus since you presuppose that God emerges simply and indistinctly from the ground, i.e., has his origin from the ground, you in fact distort my actual argument. It was clear to you from this same passage (I/7, 406) that throughout I give the ground only a relation to the existing, but not to that which is *above* the ground as it is above what exists. Only to the latter does it relate itself as the ground (groundwork) of existence. You must at least therefore add the qualification that God emerges as the existing (τὸ ὄν), as subject of existence, from out of ground to existence. But with this qualification the impossibility of the argument is clearly too immediate. For the ground of existence cannot be the ground of anything other than of *existing*, purely as such; but not the ground of *that which exists*, of the subject of existence, which, as I have already shown you, are two vastly different concepts.

[173] If you wanted to express my actual thought, you could simply say: the irrational principle in God, which for itself is without understanding, is for Him the ground, i.e. groundwork, condition, medium of revelation, of His subject which *is* only in-itself, or condition of His externally acting existence. If you wanted to dispute this thought, there are two ways to do so. [Either] you must claim that God never had to reveal Himself necessarily, that God was always already revealed, and the revealed God is one with the original essence. Or you must deny that there is a principle in God which is for itself without understanding, and that the medium of revelation could come to be, which for you according to your customary concepts could not be all that difficult.

The same relation that according to *your* presentation of my basic principles *God* has to the ground, you also seem to assume between light and darkness, understanding and the lack of understanding. You imagine, it seems, that light as *potency*, as essence, originates from darkness, and likewise the understanding as *principle* originates from the lack of understanding—what has being [*Seyende*] from what does not have being [*Nichtseyenden*]. Now here you would misunderstand me in an incomprehensible way. Darkness is not, so to speak, necessary in order for there to even be a being [*Wesen*] like light; likewise the lack of understanding does not, so to speak, need to *be* [*seyn*] for there to be understanding. Light, along with its analogue, the understanding, is a potency that is entirely independent of darkness; one could even say that it is a higher potency than darkness (=A^2). Or (to say what is actually the case), light is precisely that which has being [*das Seyende*] according to its own *concept* (in itself); *as such* it requires no other potency. But precisely, so to speak, because it is that which has its being in itself, it needs *to be* as this being which has its being in itself. That is, in order to *prove to be* this being, to reveal itself *actively* as this being, the counter-effecting principle of darkness, which is therefore the necessary ground (groundwork, foundation, condition) of its actualization, the condition by which it can be *actual*, its positing as actual, must precede it, but in so doing not in accordance with its concept but in accordance with time. Now in contrast, darkness is, as the absolute opposite of light, what necessarily does not *have being* [*das nicht* ***Seyende***] from itself (in itself), that is, it [174] has no true *being in itself by itself* [***Seyn in sich selber***] and is originally only a *being outside of itself* [***Seyn außer sich***]; for it is not so that it can be itself but rather so it can be through another, namely in order to will the light, but only as the condition and instrument of its revelation. And just as darkness is the condition or *ground* of the external-*being* [*äußerlich-**Seyn***] or existence of light, so, conversely, light is the *cause* of a being-in-itself [*in-sich-Seyn*] of darkness. Darkness is the creative principle which calls forth something that has being [*etwas Seyendes*] (the creature) out of that which does not have being [*aus dem Nichtseyenden*].

In this regard I say only that in the passage (I/7, 360) that you must have had before your eyes during the above conclusion, being *emerges* [*geht hervor*] out of nonbeing, understanding out of the lack of understanding, light out of darkness.

If you wanted, however, to make this proposition into a general one, you would need to add a proviso that accords with my thought. Understanding emerges from the lack of understanding, but only insofar as the latter is in the process of dying away, and light emerges from darkness, but only insofar as the latter has been vanquished. The understanding as *such*, as *actual*, can arise only insofar as it subjugates the prior prevailing lack of understanding and it becomes as if dead (prime matter or stuff) relative to it. Light can as such only arise into actuality when it has first killed off and subjugated to itself the initially prevailing darkness.

Thus if the earlier inference also sounds abject enough to fill everybody with horror, one need only insert the *determination* appropriate to my meaning in order to reconcile them with it. Virtue clearly does not emerge from the concept and the essence of vice, but rather in accord with the actuality in which vice has been overcome and slain. Saintliness is only possible in accord with the utter extinction of sin and emerges in deed out of sin, but insofar as it has slain it. Heaven entirely rests on hell, and this is a proposition intelligible to all. Heaven is the highest harmony, hell the discord of forces. Living harmony is overcome and subjugated discord. Heaven would be powerless without hell. There is no feeling of heaven [175] other than the constant overcoming of the hell of discord, just as there would be no feeling of health without coping with a sickness always conceived in its emergence and always again brought to silence. If God is to live in a human being, the devil in him must die, just as one could conversely say: in the human being, the devil dwells where God has departed.

As you did once already, you once again mean to engage my system, especially its ethicality [*Sittlichkeit*], which according to you has no place for duty, right, conscience, and virtue (22). It is not easy to rebut you on this claim. I could ask what moral fruits could be borne by the conjecture that the human being could only be a player on the earth, whom the sun sends and calls back as it pleases (46); or by that literary immortality (31) according to which we have the stars on high to observe as the arenas of our future metamorphoses; or from the fact that the idea of a spiritual world is entirely alien to your entire philosophy. This is demonstrated in the shared schema (21) of your system. On its negative side, the understanding has *nature*; on its positive side, *history* (also called the "higher order of things"—as if there were another!); in the middle of both is art—and that's it! Indeed, as I said previously, I do not like such inferences at all because it is self-evident that what is errant or without significance when looked at theoretically must necessarily also be insignificant or misguided when looked at morally.

I grant that virtue cannot arise from (dead) proportions and disproportions (with mathematical formulas) (22); but I assure you that this is not my intention and that you are far removed from my meaning when you attribute to me popular formulas like the "dynamic progression of degrees [*Gradreihe*], where two forces separate from a common center."

You see in my doctrine (uncontestably because of some physical expressions) that ethics has already been devoured by physics and that the higher order of things has been depotentiated to the lower (21–2). You cannot concede that the natural principles of light and gravity are also [176] applied to the ethical order of things only analogically (36). On the other hand, where you yourself wish to explain the nature of evil, you call its inner principle the deepest immoral point of *attraction* in human beings (50). How is the present connection between gravity and attraction in any way different, as I can find only my own idea in what you say of the nature of evil? "The concepts of periphery and center belong purely to the side of nature and are incapable of any meaning in the moral world" (23),[14] and indeed shortly before this passage you make nature into the negative, and the ethical world into the positive side and then compare both with the opposing curves of a hyperbola (by which, to say in passing, the ethical world is posited as completely *equal* to nature, whereas in a comparison that links the ethical with the center and nature with the periphery, at least the superiority of the moral over the physical world would be presented), and then further compare the history of humanity to a cycloid, with the claim that only the transcendent lines of infinite order serve as analogues for the ethical world (24). As if these lines were lesser forms of nature than the circle is! And if mathematical analogies for ethical things are permitted, why not the analogies of midpoint and periphery, which are far more generally meaningful than the curves, the abscises, and the asymptotes of the hyperbola that you so love.

Above all I want to note my wonderment at how you still rely on mathematical comparisons which could perhaps accord some advantages to subordinate things and how you cannot overcome the inclination to bring everything back to these dead formulas. You want nevertheless (40) to use the lever to interpret my dialectical theory of the first origin of duality, as in for instance magnetic phenomena. According to the accepted order, I should have spoken about it right at the beginning. But I forbore doing so because I found in that entire exposition (38–40) a confusion of my concepts, so that I did not know where to begin. I needed either [177] to set the entire matter right and deduce it from the beginning—who knows whether with any more success, given the silence with which you can look down on all philosophical systems with a mocking and self-deprecatory glance (41), which would have made the deduction difficult for you to follow. Or I needed to emulate that place in your letter where you, in order to pronounce yourself fully (24–5), speak in loud negations. I needed to say, "A living process, like that of the first origin of duality, cannot be presented as a lever. What I call the ground cannot be compared with the center of gravity (40). And if such a mechanization were called for here, the ground would have to be compared with one weight of a lever. There does not have to be a particular differentiating principle in the one in order to explain the origin of duality (38). I have never maintained that evil and good—simultaneously or successively—emerge from the nonground" (39), etc. But what purpose do all of these denials serve? The course of the historical and genetic process is unfamiliar to you, so how could you, or anyone else like you, rightly find your way through the dialectical process?

With this I would thus like to conclude my answers to your objections in that I believe that what has been presented up to now is enough to convince you that you have not yet fully grasped the sense and connection of the ideas contained in my essay. I pass over much that is easily soluble or thoroughly errant, such as when you pronounce (25) me guilty of denying the freedom of God because I explain the bond of the principles in him as (morally) irresolvable, or when you impose beginningless time (42) onto Nature Philosophy—(frankly I do not know what you mean by this).

If to this point I have perhaps been a little too curt and dry in answering your objections, it is befitting for me now to confess that these objections, like your views of my system in general, follow from your own way of viewing things, and that the place where you stand cannot appear otherwise to you. My answer would thus be incomplete and insufficient if [178] you did not allow me now to share my own thoughts on your philosophical standpoint.

"Creation does not come into being, but rather is given. As *human reason* is posited, so also is the idea of truth posited in it, and creation or (?) nature is nothing other than the total reflection of the idea of truth" (43). It does not come immediately in any way from God, but the so-called idea, *truth*, is its own creator (which is the same). This so-called idea, however, is itself only posited insofar as human reason is posited, and simultaneously with it. Thus for you all of nature falls within human reason. This view thus does not raise itself over the most utterly subjective idealism, but it is not the living ego, as with Fichte, but, in its mediated form, cold reason, and immediately the dead concept of truth by which nature produces, or, since this is already something living, is simply posited. I find this type of idealism the most convenient of all. It throws out all questions concerning origin or *happening*, which are always the most difficult matters. The question if and when nature can arise has no meaning because the idea of truth, the soul of all constructions, cannot be taken back up into the construction. There are people who on account of an inborn insensitivity to color view the world as if it were a copper etching. People for whom nature would become *actually* what it is in your thought, a total reflection of the so-called idea of freedom, must experience it doubtlessly as a book full of geometrical figures. The word reflection would, by the way, make the matter entirely too simple if only a reflection could be thought without something from which or out of which it is reflected. But since you place nothing outside of human reason and the so-called idea of truth posited therein, I am eager to know where you would like to obtain the medium that would reflect that totality.

As it was on the side of nature, so it is with the ethical side: "world-history is nothing other than the total reflection of the *idea of virtue*[''] (45). It would [179] do the harried human race well to be granted such a simple and innocent world-history. But a history in which you yourself can very easily observe the appearances and effects of the devil if you were only willing to do him this honor (50b), in which all forces of evil have shown themselves openly alongside rare and concealed virtues—to want to grasp such a history as a total reflection of virtue, seems to me far too philanthropic.

[''']Likewise the organism is only the total reflection of the idea of beauty and given together with reason[''] (47). If the charmed observer avows that all individual beauties of the universe play together in the most delightful harmony upon the surface of the perfect human body, no one would say [the same about] the stomach, intestines, spleen, liver, pancreas, and the many individual organs protecting against nature which have been brought into the interior and also belong to the organism, or that the organism that is responsible for the human being's feeling of beauty and yet fills so many creatures with disgust is a total reflection of beauty.

Once again: I find a reason *with which everything is already posited* uncommonly convenient and only wonder how you still find reflections of those three abstractions necessary. It would be far briefer to posit nature, history, and organism (which you seem not to count as [a part of] nature) without such qualifications as equal in their entire breadth with reason and especially with its three abstractions.

You are uncontestably astounded by the expression I use regarding these so-called ideas. For it is already certain and decided, and induces on a daily basis the most beautiful speech, that truth, virtue, and beauty are ideas. And this is understood quite literally. It is also held by true Platonists. For Plato says for example in the *Phaedrus* of *beauty*, that we see it first in the highest glance and recall it when we here view beautiful things. To the extent I can presume to understand Plato's meaning, he means here the original beauty—not [180] what-has-being-for-itself [*für-sich-seyende*] since in itself it cannot be a subject, but the beauty *of what is in and from itself beautiful*, and this, as an actual *subject* (an ὄντος ὄν), can alone be called a *primordial image* [*Ur-Bild*], an *idea* according to the Platonic use of language and manner of thinking. It is also well known that Plato does not speak of *equality*, of *beauty*, *goodness*, *justice*, *saintliness*, but of the *equal itself* (αὔτὸ τὸ ἴσον, *Phaedo*, 74a),[15] of *beauty*, of *goodness*, of *justice*, of *saintliness* themselves (περὶ αὔτοῦ τοῦ καλοῦ, καὶ αὔτοῦ τοῦ ἀγαθοῦ καὶ δικαίου υκαί ὁσίου, *Phaedo*, 75d). Now it is frankly true that he who sees the good in itself, the beautiful from itself, the equal in itself to itself, also views the primordial image of *beauty*, the primordial image of *goodness* and *equality*. But the primordial image of these things is not beauty as a property or as an abstraction, but rather what has being [*das Seyende*] (τὸ ὄν), the subject, primordial beauty [*das Ur-Schöne*] itself.

Now it is precisely what is most proper to your philosophy, or, as you yourself prefer to call it, your non-philosophy, that above all it wants no ὄν, no being truly and in itself because God would then have to become particular, a subject. Reason is for you the universal organ; it has for its immediate object the pure totality (7). The eternal is for you not one, but the *all* (26). The outrage of the vexing subject consists (44) in it reducing totality to unity, or, as you explain it, in that it wants to make everything particular, individual, personal—into a subject or being [*Seyenden*]. Unity is life and the soul in their totality. A reduction of totality to unity thus seems to me nothing other than a *reduction* of the living body [*Leibes*] to the *soul*, of the body [*Körpers*] to the *spirit*.

The abstractions truth, virtue, and beauty are well suited to such a self- and subject-less totality, because they do not at all need to assume a subject or to *reduce* totality to unity; and the absolute may only *lie* for you in the *harmony* of these so-called ideas [181] (what do you mean by *this* word [i.e., 'ideas'], or with the word *lie*?), which, as I myself here freely see, is nothing divine. For three abstractions, to which an equally abstract harmony is inferior, are not even close to forming a subject, let alone a divine one. It is simply astonishing that you still speak of an absolute (why not of a mere absoluteness?).

You yourself freely acknowledge that for you so-called reason does not behave anything like an abstraction; thus you flee the divine to a realm that lies far above all reason and understanding, in which freedom and virtue themselves vanish as dark points (48b). But in doing so you in fact gain nothing. For if you too have now gotten far beyond reason in your thoughts, the *veto* of this universal organ follows you, with the result that God may not be anything particular, i.e. personal, a subject, and at the same time you remain gripped by a fear of the understanding, which in your mind wants to kill everything that originally belongs only to totality (18). And the God that you seek there may only be one of whom no predicates, not even omnipotence, beneficence, or omniscience, may be spoken. He is thus subjectless, a God which does not *have being* [*nicht seyender Gott*]. You are so captive that even in the flight from it you are held down by the pressure of its bonds and universal abstractions.

If I must conjecture, you seek our scientific difference in my reluctance to follow your flight into a higher realm or even an inclination to blame you for it. Nothing less! I only wished that you rid yourself of that abstract reason. True, from the very beginning you announce its inefficacy not only where God is at issue, for it is equally inept at understanding nature, life or history. But insofar as you believe yourself to have completely set yourself free of this reason and its realm, I find you right in the middle of it. For how else should we take your frequent expressions that [182] God may not be personal, etc. other than by the power decree of your universal organ, a universal that here means against everything that has being [*alles Seyende*] and hostile to anything that is or is called subject?

May I allow myself to enlighten you once again regarding this self-deception in which you appear to me to be caught? You call reason the universal organ, in which totality exclusively resides (7).[16] [You also call] the understanding, in contrast, the particular organ, in which unity [*Einheit*] dominates and seeks to particularize everything. Now it is the particular that is the living and the self- or subject-less totality that is the nonliving, just as the *beautiful*, the good, the true, are living, but truth, goodness, and beauty as such are non-living. Thus according to your own assertion it is reason which kills and the understanding which enlivens. Or if as you say in another passage the concept and the understanding at work in it kill what is living, it is precisely what you call reason which posits the totality over the unity (the dead over the living), and while you rage against the understanding, you find yourself in the all-destructive violence of the understanding, which tolerates the still abstract concept but nothing that *is* or lives. Thus as we can often observe that people most fiercely resist just what biases them most, one often only needs to take care to suppose that those things about which one expresses oneself most heatedly are the ones by which one is most controlled.

In fact, whereas you disparage the *concept* so severely, I would know on my side not to employ it lightly and thereby conceive a dead philosophy of the concept, a lifeless formalism, in a system constructed from the reflections of truth, virtue, beauty, and the combination of such abstractions.

Under these circumstances it is a credit to your feelings when you (29) assure that we also cannot formulate one idea of God. I find this expression nothing less than hyperbolic, and if not directly beneath God's dignity, then [183] still inappropriate for the poverty of tools that you ascribe to human beings.—Specifically I could never find God's dignity enhanced if it were not possible for the most perfect of God's creations to form a thought of Him. Indeed, you posit the human being, whose race (the youngest of all) could be viewed as eternal according to another passage, as intellectually far beneath your (not yet so stipulated) extraterrestrials. But since this occurs without reason [*ohne Gründe*], you will in the meantime otherwise permit us to maintain our faith that the human being receives a very high position in the general ladder of rational beings. That the earth swims in the uncountable number of stars only as a small point (45) is no argument against this, since God does not make distinctions of greatness or superiority in caring for his gifts, and according to a basic law of his household (known to every deep scholar), which is to be reached only in grasping our enigmatic relation, rather prefers the humble and base. You scoff that it *falls* to us to make ourselves into the image [*Ebenbild*] of God,[17] to which the understanding also adds its two cents, in that it shows quite artificially how God was actually *forced* to create such a corporeal image of himself (30). My belief in contrast is that it did not fall to humans to become the image of God, but rather that God himself made the human being in his image, against which it was certainly a different and opposed *Fall* (a Fall of human beings and the devil) by which the human being became the non-image of God.[18] In this devaluation of the human being (also of the original), your philosophy agrees completely, like all new doctrines that boil down to nescience [*Nichtwissen*], with the system of the most corrupt Enlightenment, whose purpose is none other than to separate human beings from God as much as possible—a hidden agreement to which one cannot be too attentive.

To this end I recall the passage in Malebranche's preface to his work *Recherche de la Verité*, which reads:

> [184] The mind [*Geist*] of man is by its nature situated, as it were, between its Creator and corporeal creatures, for, according to Saint Augustine, there is nothing but God above it and nothing but bodies below it. But as the mind's position above all material things does not prevent it from being joined to them, and even depending in way on part of matter, so the infinite distance between the sovereign Being [*menschliches Wesen*] and the mind of man [*menschlicher Geist*] does not prevent it from being immediately joined to it in a very intimate way . . . I am not surprised that ordinary men or pagan philosophers consider only the soul's [*Geist*] relation and union with the body, without recognizing the relation and union it has with God; but I am surprised that Christian philosophers, who ought to prefer the mind of God to the mind of man, Moses to Aristotle, and Saint Augustine to some worthless commentator on a pagan philosopher, should regard [*in solchem Irrthum schweben*] the soul [*Geist*] more as the *form* of the body than as being made in the image and for the image of God, i.e., according to Saint Augustine, for the Truth to which alone it is immediately joined.[19]

You will uncontestably reply to this that this unification of the human being and God which you overturn [*aufheben*] in cognition is reestablished through faith, to which you wish to assign exclusively all properties of a forceful conviction originating out of the entire human being—which follows from what you previously pronounced as the capacities of the understanding and reason. But I ask you how a faith that cannot grasp even a single idea of God, that does not permit us to say that He is, let alone that He is personal, self-conscious, spiritual, omnibenevolent—how such a thoughtless, spiritually dead, dumb faith can carry out a connection of the human spirit with God. Indeed, faith is not possible in *that* sense, and a faith with which no thought and no knowledge [185] is connected cancels itself [*hebt sich selbst auf*]. Explain this as you will, or do not explain it; for faith is a function of what is innermost in us that is inseparable from consciousness, and thus from knowing and thinking. It is well known that the word faith comes to us from sacred writings and most immediately from our Lutheran concept and actually expresses only confidence of conviction, the unanimity of the heart with certain cognition. Genuine faith is itself nothing other than a faithful, that is, confident knowing, in which, as in *all* true knowledge, heart and spirit are in harmony. But in no way is it, as you and some others wish, a complete negation of all knowing.

Declarations in which you—however rigorously you apply your own concepts—deny all consolation of knowing to the human being even in view of everything individual, whose cognition and correct assessment is necessary to one's salvation and reassurance; declarations such as your claim (49) that the question as to why there is evil in the world is not only nonsense, but impudent, are, to my reckoning, contrary to the sense of a God-given faith. If God *wanted* to conceal from us the ground of the allowance [*Zulassung*] of evil, it is uncontestably not only presumptuous but also insane for us to want to know it. But how do you *know* this from your own principles? If, however, God actually wants us as co-knowers[20] of his eternal plan, what—to turn your question (12) back to you—do you have against this? Thus it is eternally not only permitted, but commanded of the human being that he seek to know God's intentions, even the most concealed ones, but not to remain far and alien like a mere house slave, but to become familiar and at home in the kingdom of the Father, like the Son whom he has determined as the Trustee of his mysteries and educates ever after. This is the great sense of what the doubters and agnostics of Christianity do not even have an inkling about: that it has torn down the dividing wall, that we now may approach and, as the Apostle says, all have access in one spirit to the father.[21]

[186] You speak very intensively of Christianity. But partly it is in fact not so simple (52) as you would like to present it, namely as a doctrine simply speaking to the heart, revealing no mysteries to the spirit; and partly I do not know if Christianity's *actual simplicity*, e.g. the doctrine that God sent his son, can be reconciled with the sublimity of *your* concepts of God and his dignity. For how should a God whom only a laughable understanding makes into a superlative of human nature come into such human relations that he has a son whom he sends? Or how should this "small point that is earth, which can with its political fragments arise and pass away without the *world-history of the universe* being in the least disturbed" (45) have been destined to be the primary arena of God's revelations? Or how could the condition of the human being who deserves the same wan smile of your "inhabitants of the sun" when he *gloats* of being in the image of God that we would give to the little worms in the grass if they made themselves into the image of human beings (31), and the course of human history, which is perhaps only a novel or drama that the sun has written, have been so important in the eyes of Him who is above all suns?

Frankly you reduce this meaning of Christianity again to a fairly accessible level in repeatedly calling God's revelation through Christianity a mediated one (51, 52, etc.), and retain the word according to the favored means of disclosure of enlightened theologians, but cancel (*aufheben*) the force of the matter. At least I do not thus conceive how you call the attempt to explain natural wonders irreligious, since with a natural explanation of wonders, of Christ, for example, their mediated-divine origin can very well exist.

But not only the concept of revelation, [but] all other concepts of Christianity must consequently for you be such mediated, that is, improper, concepts. If it must be called offensive to pure representations to think God with human properties and it is only falls to humans [*Einfall*] to think the human being as [187] in the original image of God, then the same must be said of believing that the human being can recuperate this image. If this is true in all cases, then the whole of Christianity is an expressionless, cold and, as such, tasteless allegory which one had better lead back to its *simple*, imageless [*unbildlich*], sober meaning. The striving really to attempt to do this deserves to be called religious rather than irreligious in your sense. For this striving arises, at least in its better [forms], from the same *purity* of representations of God which you exhort all human concepts of God to reject.

I would incidentally be thrilled if you should find in the *purity* of these views a means of navigating the decay of time and allow the same justice to befall the depth of your sensitivity—although I do not see how our character and convictions should worsen (32) under the influence of ***'s text book any more than they could improve in the process. You accurately present how far it has come with our time (54). But against the effectiveness of the means I allow myself only a single doubt. The sermon of a faith opposed to all knowledge (which, as it appears to me, begins ultimately in the root of the unbelief in one), was in the world long before us and before the events of the last decade. And its enunciators, like Jacobi, certainly had the best wills—but even Kant led in this direction. And indeed it had proved itself completely ineffectual and it had not saved in the slightest (not to say any more). How should the drug which could not prevent the ill of general unbelief and the even worse ill of indifference heal the violence that had already entered in its entirety? Which curative permits itself in the least to expect a philosophy that consists in loud negations of everything concerning what is higher and pulls itself back from the most important questions for human beings into nescience? For this nescience of the higher can also facilitate that nefarious brand of "Enlightenment" [*Aufklärei*], which is hostile to everything higher and which would be overjoyed to see the attainment [188] of its main goal of a visibly abject, even morally bankrupt, way.

When Christ appeared in the time of greatest corruption, he did not on that account begin to say: We can know nothing of God, formulate no thoughts of him; still less can it occur to us [*einfallen lassen*] to be in his image. Even the humanity of our time yearns for something positive that can reflect only a more forceful understanding, one more powerful than the true ideas. Then the wise will again believe in a God, as God was our father, which no one can *properly* do today. In this respect I have especially considered your correspondence as one of the most remarkable documents of the spirit of our time.

And thus this long answer has at last come to an end. If I may briefly summarize *our* relationship, it seems to stand like so. You lack the actual intermediary concepts [*Mittelbegriffe*][22] of my system. You know my words and individual sentences, which you then understand in your own way. The chief reason for this is that you have not yet detached yourself from Fichteanism and appear to hold my view only for a further development of his and thus also interpret my doctrines according to the concepts you have brought over from his system. For me what is most distressing is your belief that you have actually understood me, and that you perhaps interpret my assurance to the contrary as arrogance. Just as I did not devise my system of thought in a single day, my views in their entire connection also cannot be grasped in a single day. It seems impossible to me that the system is so easy to dispose of that you could have looked over and grasped the whole from your customary chair. The basis on which you find yourself cannot support this whole. Thus it cannot come *to you*, but you must come to it, just as one who wants to view the cathedral at Strasbourg (beyond compare, let it be noted!) must separate from his seat and go there because it cannot stir from its place and come to him. This says nothing of the value of your [189] standpoint. You may consider it the best, as is fair, and also return to it now that you have shown your hand. If, however, you want to look into my system and judge it, you would have to completely distance yourself from your own standpoint and import nothing of what it takes to be true.

In considering your own doctrines or convictions, I lament that a man of your spirit fixes his goal so early in his investigation and entrusts too soon certain concepts that could not pass a more careful examination. For example, what you say about understanding and reason is all taken from a web of thought into which not a ray of critique has shone, as it must not have even begun to shed the light of proper philosophy onto what reason and understanding are, instead illuminating these with arbitrarily assumed concepts. Thus I wish very much that it would for once please you, instead of arguing from mere presuppositions, to go back to what is really prime [*wirklich Erste*] and above all to sift critically through the entire inventory of concepts which you take for generally accepted currency.

Now if I have said all of this too directly, laid it down too nakedly and purely, then excuse me with the vital wish that with this occasion no doubt is left between us, and let us one time come to an understanding across the opposed sides of our relationship. When it comes to truth, personal esteem ceases to matter; but what I have put forward is also in no way directed against you, whose true essence and willing I should do well to distinguish from individual utterances, but rather against the entire manner of thinking which you would presume to defend.

And thus live well.

APPENDIX B

Schelling's Unfinished Dialogue: Reason and Personality in the Letter to Eschenmayer

CHRISTOPHER LAUER

Schelling opens his most famous letter to Eschenmayer, first published in 1812 in the *Allgemeine Zeitschrift von Deutschen für Deutschen*, by asking that it be read as a dialogue. And while the letter's dismissiveness and occasional nastiness[1] would seem to suggest that he has other priorities than an open exchange of ideas, it is quite possible to take him at his word. Despite the possessive or even territorial sound of such Schellingian titles as *Presentation of My System of Philosophy* and *Further Presentations of My System of Philosophy*, Schelling never took his work to be the final and pure expression of reason, and he voices doubt in the *Freedom* essay that "a system, as it is composed in the head of any one individual, is ever the system of reason κατ' ἐξοχήν, the eternally unchangeable" (I/7, 347).[2] Thus it should not be surprising that some of Schelling's clearest and most vivid expressions of his philosophy of freedom emerge in his response to a colleague. Certainly, Schelling was highly ambivalent about interacting with his contemporaries, even scornfully citing in the *Freedom* essay Erasmus's "*semper solus esse volui nihilque pejus odi quam juratos et factiosos*"—"I always wanted to be alone, and nothing more did I hate than conspirators and factionists" (I/7, 409n). Nevertheless, this letter from one of Schelling's colleagues and most avid (if not faithful) readers was such a welcome opportunity to develop his ideas that Schelling, according to Xavier Tilliette's biography, viewed it as "a gift from heaven."[3] Though he had from time to time toyed with dialogical formats in his earlier works, including most notably his *Philosophical Letters on Dogmatism and Criticism* (1795) and *Bruno, or On the Natural and the Divine Principle of Things* (1802), he judged the *Freedom* essay a new sort of dialogue in which the human spirit could converse not only with itself, but with nature (I/7, 333). He thus laments in the same footnote in which he echoes Erasmus's disdain that he had not been able to match the essay's underlying movement, with which "everything arises as a sort of dialogue," with a properly dialogical form. With this aim of prolonging this dialogue, in this essay I want to trace out two important ways in which Schelling's dialogue with Eschenmayer expands on the sense of the *Freedom* essay and then point to a third discussion in which an opportunity for further elucidation was missed.

Existence and Ground

The most famous product of Eschenmayer and Schelling's exchange is the discussion of God's duality as existing personality and as ground of existence. Eschenmayer charges that Schelling has misconceived God's eternity by applying to him a distinction between ground and existence that is only valid for God's creatures. Indeed, he argues, it is nonsensical to speak of God having a ground "in Himself," because "'in-Himself' and 'outside-Himself' have no meaning for God; there is no ground that acts independently of God, which for you contains the possibility of the evil principle" (I/8, 150).[4] God's existence is of a different character than human beings' and thus can only be one and the same with His ground.

Schelling's pleasure in the opportunity this affords him to clarify a crucial point in the *Freedom* essay is evident in the uncharacteristic magnanimity with which he begins his answer to Eschenmayer. The concept of *existence*, he admits, is ambiguous, and thus Eschenmayer could indeed be right when he argues that existence and ground of existence must collapse into one (*in eins zusammenfallen*). Provided we understand God's existence as His *mere existing*, that is, as the simple fact that He is, then it is perfectly acceptable to assert this identity. But if we wish to speak of God's totality as a personality, then His being as ground and as existing must be distinct.

By insisting on the former concept of existence, it is actually Eschenmayer who reduces God to a mere object of the understanding (*Verstandeswesen*). For Eschenmayer assumes that the only possible relation of a ground to an existing being is that of antecedent to a consequence. Granted, God as antecedent and as consequence cannot be distinct, because the understanding has always conceived God as *causa sui*. But in the *Freedom* essay Schelling develops a much richer conception of ground as organic, which, as we see below, is not easily brought under the control of what the German idealists called the understanding (*Verstand*). Eschenmayer was by no means alone in assuming this abstract conception of ground, as Schelling famously claims that "the whole of modern European philosophy since its beginning (with Descartes) has the common defect that nature is not present for it and that it lacks a living ground" (I/7, 356).[5]

As Schelling clarifies in the letter to Eschenmayer, this means that the ground is precisely the not-being-itself that a living personality requires (I/8, 164). In Leibnizian terms, nature, or God's ground, is what outstrips the divine will's ability to actualize what the divine understanding judges the best of all possible worlds. While God in his existing wills only the good and hence is completely identical with himself, the ground from out of which he wills the good lies beyond this simple act of self-willing. Yet the fact that God could not bring himself into a personal existence without a ground does not imply that the ground produces God any more than the fact that all light is grounded in darkness implies that darkness produces light (I/8, 173). Light in Schelling's Nature Philosophy is nature's principle of self-formation or being-for-itself, and thus it is more accurate to say that while the dark ground precedes light chronologically, it is only light that has sufficient being to allow darkness to distinguish itself as a principle. For Schelling in the *Freedom* essay, selfhood, or personality in general (and not just for the human being), depends on an endeavor not simply to persist, but to produce (*herstellen*) oneself, which entails separating oneself from one's ground. The peculiar structure of human freedom is not yet at issue here, and Schelling's claim about the need for God to lift himself out of his ground is independent of the complex interplay of the dark and light principles in the human being. It is not merely human freedom that would be impossible without a divine personality, but nature in general, which must be determined in some positive way if it is not to collapse into mere sameness (*Einerleiheit*) with God (I/7, 344).

Schelling admits that any system must account for the original unity of God with this ground beyond the divine will if it is to avoid the self-laceration of Manichaean dualism (I/7, 406n). In the letter to Eschenmayer, his word for this original unity (which does not appear in this technical sense in the *Freedom* essay)[6] is *Urwesen*, primordial essence.[7] While Schelling does mean to distinguish this essence from God's existence as ground and as self, he does not mean it in a neoplatonic sense where it would determine God's existence or be the ground from which these existences emanate. Rather, it is, as Schelling puts it in the *Freedom* essay, a nonground (*Ungrund*) indifferent to God's ground and selfhood (I/7, 406). Though the nonground is ontologically prior to ground and existence, it is not the ground of their identity, but the very possibility of their difference. As such, Eschenmayer's claim that God's ground and existence "collapse into one" has missed perhaps the most important insight of the *Freedom* essay.

Reason, Understanding, and the Thing

If the most famous product of the exchange with Eschenmayer is the chance it affords Schelling to expand a central *topos* in the *Freedom* essay, one of its underappreciated products is the opportunity it gives Schelling to *refuse* to expand on his terminology in the essay. Twice Schelling almost lets Eschenmayer lure him into debates over the relative merits of reason and the understanding, but each time Schelling manages to abstain by insisting, "I do not want to get involved in this disputed question" (I/8, 166, 167). While the terms "reason" (*Vernunft*) and "understanding" (*Verstand*) appear throughout the *Freedom* essay,[8] Schelling never bothers to define them. Given the tremendous effort Schelling expended in the first decade of his publishing career attempting to redefine these crucial Kantian terms, one might suspect that he simply felt no need to define them in the *Freedom* essay. In his 1797/1803 *Ideas for a Philosophy of Nature*, for instance, Schelling works to show that while philosophy begins with the obsessive categorization of the understanding (which Eschenmayer here calls particularization), it can only complete its project by stepping beyond this obsession into reason's drive for unity.[9] And yet in the *Freedom* essay, this distinction has been suspended, so that the irrational core from which reason emerges can just as well be called *das Verstandlose*, that which is without understanding. Since the capacity (*Vermögen*) to choose between good and evil is philosophy's central concern, making distinctions between the other faculties has become a peripheral matter. Thus when Eschenmayer speaks of reason and understanding in radically different (and possibly self-contradictory) ways than Schelling, Schelling does not take the opportunity the dialogue affords him to "correct misunderstandings as they arise," but claims that these matters are beside the point. Specifically, he argues that discussions of whether the various human faculties of knowledge are adequate to God's perfection are secondary to the question of what God *must* be like if there is to be such a thing as nature at all (I/8, 167).

Thus by bowing out of an easily winnable battle, Schelling advances the goals of the *Freedom* essay far more powerfully than he would have by responding to Eschenmayer's provocation. In the preface attached to the *Freedom* essay at its first publication, Schelling boldly asserts that it was his intention to remove "reason, thinking, and knowledge" from their established place at the core of philosophical inquiry (I/7, 333). In its obsession with a particularly human knowledge, philosophy had elevated these cognitive activities above all others in establishing a division between rational spirit and mechanical nature. But now that "that root of opposition has been torn up" and its humanistic vision proved an effete fantasy, "it is time for the higher, or rather the actual [*eigentliche*] opposition to come to the fore, the opposition of necessity and freedom, with which the innermost center of philosophy first comes into view" (I/7, 333).

Eschenmayer, however, wants to move in the opposite direction, by positing that the ideas of reason (truth, beauty, and goodness), though still inadequate to God's perfection, are on account of their universal immediacy the best approximation or reflection we have of the divine.[10] While Schelling dismisses this a priori access to the divine as "uncommonly convenient" (I/8, 179), it is close enough to some of Schelling's earlier work[11] that it is worth elaborating in detail why it is incompatible with the movement of the *Freedom* essay. In asking whether Schelling's analysis of God's structure reduces God himself to an object of the understanding (*Verstandeswesen*), Eschenmayer recalls Schelling's distinction in the *Presentation of My System of Philosophy* between the analytic understanding on the one hand and reason on the other, "which does not have the idea of God, but is this idea itself and nothing else besides" (I/8, 146). In order to think God's separation into ground and existence, he argues, we would have to abandon reason's most basic knowledge of God's unity, releasing the understanding from the only moorings that could possibly ground it.

Schelling agrees fully, but counters that these moorings are themselves unthinkable. In elevating reason over the understanding, the prior works of German idealism—including Schelling's own works—hold that the understanding can have no access to the will's (including God's will's) tendency to return to itself, which the understanding can view only as part of an abyss of ungrounded desires. Whereas reason finds its truth in willing, the understanding obsessively dissects all desires in the vain hope of finding some truth beneath them. Thus it can only be regarded as a deep betrayal of Schelling's idealistic roots when he responds to Eschenmayer's criticisms by asking, "Can a being [*Wesen*] acting in accord with a purpose and intention even be thought that is not *eo ipso* also an object of the understanding [*Verstandeswesen*]?" (I/8, 166). Here Schelling is questioning not only the absolute priority of reason over the understanding, but also German idealism's tradition of privileging the intelligibility given by reason. The early Schelling had argued that it was reason, and not the understanding, that was key to conceiving both the teleological and self-organizing structure of organic nature and the freedom of thought.[12] Accordingly, he argued that the self-sufficiency of organic beings was inaccessible to the understanding, which ceaselessly analyzes any object it encounters, ignoring the wholeness that living beings give themselves by positing their own ends. For Schelling to grant the understanding a place in cognizing willing and to think of it as itself a self-subsistent form of striving is to deny reason the right to distinguish itself absolutely from the understanding. By locating reason in relation to its ground rather than to its self-genesis, Schelling places reason together with the understanding along the periphery of philosophy.

The distinction between reason and the understanding is thus not to be conceived solely by reason, but instead by the freedom that makes both possible. What Eschenmayer proposes in place of such a living understanding Schelling calls "cold reason, and immediately the dead concept of truth . . . I find this type of idealism the most convenient [*bequemste*] of all. It throws out all questions of origin, of *happening*, which are always the most difficult matters" (I/8, 178). If philosophy is to provide any clue regarding the origin of human freedom, it must not dissolve the universe into such a simple identity. For under such a totalized reason, the individual human being could only be a reflection (*Reflex*) of the idea and would have no ability to depart from divine providence through sin (I/8, 178).

But rather than working out the precise place of the understanding in a new philosophy that places reason at the periphery, Schelling offers a mere glimpse. The original longing at the ground of God "directs itself toward the understanding, which it does not yet know, as we in our longing yearn for [*in der Sehnsucht . . . verlangen*] an unknown, nameless good, and it moves presentiently like an undulating, surging sea, similar to Plato's matter, following a dark, uncertain law, incapable of forming something lasting by itself" (I/7, 360). The possibility of something lasting, of a constant presence, only arrives on the scene when God, roused to existence, beholds himself in an image. This image, which Schelling also calls a representation (*Vorstellung*), is the *word* of God's longing, "in the sense that we say the word of [i.e., the solution to] a riddle" (I/7, 361n). In this word or answer, we find the wholeness of reason's ability to return to itself through self-reflection. Through this wholeness, the divine understanding is finally born and begins to impose its will on previously anarchic nature: "The first effect of the understanding in nature is the division of forces, since only thereby is the understanding able to unfold the unity which is contained unconsciously but necessarily in nature as in a seed" (I/7, 361). Organic life comes into being when the pure light of the understanding reaches into its dark ground and violently disrupts its undulating repose, giving it the powers of self-formation, disease, and death. Such quickening is possible because "this essence (initial nature) is nothing other than the eternal ground of God's existence, and therefore it must contain within itself, although locked up, the essence of God as a gleaming spark of life in the darkness of the deep" (I/7, 361). In the same way that sickness is only possible in health and health only apprehensible in relation to sickness, it is only with the light of divine revelation that the dark ground's potential begins to appear and only over against this ground that light is intelligible.[13]

Yet there are limits to this illumination and intelligibility. For "longing, aroused by the understanding, strives henceforth to contain the spark of life apprehended within itself, and to lock itself within itself in order that a ground always remain" (I/7, 361). Thus while Schelling would here probably be willing to concede Hegel's claim in the preface to the *Phenomenology* that the understanding is the "most astonishing and mightiest of powers," he could in no way accept the inflated claim that it is also "the absolute power."[14] For evil (and therefore freedom) is conceivable within a system only so long as the divine understanding is not absolute, only so long as it fails to spread the light of divine goodness throughout the universe. The understanding thus plays a vital, yet nevertheless secondary and only partially defined role in the *Freedom* essay. Although it helps to mark the distance between divine goodness and the ground that makes this goodness as well as evil possible, it yields no firm determinations that would allow human freedom to be *understood.*

Yet even in his justifiable refusal to get involved in a discussion he no longer sees as central to philosophy, Schelling gives the tantalizing impression that he would like to do much more to knock reason off the high perch on which the German idealists have placed it. Allowing himself a brief paragraph to consider the implications of Eschenmayer's faculty epistemology, Schelling notes that if reason is taken to be thought's organ of universality and the understanding its organ of particularity, then it is the understanding that grasps nature's living differentiation and reason that suffocates nature in lifeless abstractions (I/8, 182). We should not read too much into this disparaging of reason, for Schelling's early triumphalism on behalf of reason was not based on nearly so simplistic a distinction between reason and the understanding.[15] But still, taken together with the claim in the *Freedom* essay that once it has been generated in the soul, "a thought is an independent power, continuing to act on its own, indeed, growing within the human soul in such a way that it restrains and subjugates its own mother" (I/7, 347), the passage seems to foreshadow Schelling's later claim that reason is not merely distinct from human freedom, but is actually a disease infecting it.[16] When he observes that "even the humanity of our time yearns for something positive that can reflect only a more forceful understanding, one more powerful than the true ideas" (I/8, 188), he seems to be pointing to the positive philosophy's displacement of reason in the cognition of actual experience.

Physics, Metaphor, and Anthropomorphism

But if the aims of the *Freedom* essay justify Schelling's refusal to elaborate on these matters, they do not excuse his failure to account for his various metaphors linking nature and human moral action. Unfortunately, much like many of the Platonic dialogues Schelling wishes his confrontation with Eschenmayer to emulate, the discussion breaks down just where it could have proved most useful to "an audience of learned witnesses." When Eschenmayer accuses Schelling of allowing ethics to be swallowed by physics (I/8, 150), Schelling is quite right to cry hypocrisy, because Eschenmayer himself appeals to mathematical concepts of the understanding (such as the cycloid) to explain human moral history. Schelling's testiness is understandable, since his intention is clearly not to explain human morality by physical laws. In fact, by insisting that every moral judgment has been decided from eternity, Schelling affirms the unfathomable gulf that Kant saw between the concepts of nature and freedom.[17] Yet the shallowness of this attack allows Schelling to skirt a much more difficult issue: how analogies between the natural and moral realms can be presented in a philosophical system.

In his 1936 lecture course on the *Freedom* essay, Heidegger argues that one of the central axes on which the essay breaks down is in showing how human concepts are applied to nature and natural ones to human beings. To be fair, Heidegger remarks, the objection that the use of such human terms as "longing" to describe God is improperly anthropomorphic misses the point because one of Schelling's central aims in identifying the essence of human freedom is reevaluating what place the human can have in a philosophical system.[18] In fact, Heidegger argues, such objections only entrench anthropomorphism in philosophy by assuming that we already know the proper relationship of the human being and the world and thus can dismiss reconsiderations of this relationship out of hand. The real objection to the *Freedom* essay for Heidegger is that it fails to specify the ground from which human beings can question the ground of their freedom. If the nature of the anthropic is put in question by Schelling's anthropomorphisms, Heidegger is ultimately asking, then what is the form of this question, and how can a system of philosophy allow it to be asked?

Conversely, when Schelling states that there is an "analogy" between the relationships of ground and existence and gravity and light, "this does not mean that it is only a pictorial image. Rather, he means a justified comparison of the one stage of being with the other, both identical in the essence of being and different only in potency" (ST, 115). If philosophy is permitted to speak of the human being as an image of God and the ground as an analogue of gravity, and if the faculty of reason is no longer solely responsible for showing the identity of modes within a system of philosophy, then Eschenmayer is right to demand a fuller discussion of how Schelling purports to speak of the ground. Though he does not explain why this is the case, Heidegger is right that this requires a reconsideration of the structure of nature's potencies, for gravity and light can be analogous to ground and existence only if nature emerges not simply through the inhibition of the absolute as in Schelling's early philosophy of nature,[19] but through God's free act of expression. And because Schelling was hard at work on just such a reconsideration of the potencies for his *Ages of the World*, the exchange with Eschenmayer could have been much more illuminating if Schelling had taken this objection more seriously than the form of Eschenmayer's objection required.

If the dialogue could be prolonged, one would like an Eleatic stranger to step in and continue the tiring Eschenmayer's elenchus. Since we can no longer avoid a lifeless formalism by privileging reason over the understanding, this stranger might ask, is philosophy to maintain its vitality through a new manner of presentation? Schelling's decision to publish the *Freedom* essay without giving it the dialogical form he claims it deserved would seem to argue against this possibility, but Schelling's statements to Eschenmayer begin to make it seem plausible. After "enlightening" Eschenmayer regarding the totalizing and morbid results of elevating reason in philosophy, Schelling warns him that "we can often observe that people most fiercely resist just what biases them most" (I/8, 182), which implies that it is Eschenmayer's very tendency to formalism that leads him to impugn the understanding so harshly. On the other hand, because Schelling himself understands the vitality of nature, he is not driven unconsciously into a frenzy to eliminate formalism (I/8, 182). This, he claims, helps him avoid a consequence he draws from Eschenmayer's insistence that human concepts are inadequate to God's divinity. For if all religious thought must be purified of anthropomorphism, "then the whole of Christianity is an expressionless, cold, tasteless allegory which one must sooner lead back to its *simple*, imageless, sober meaning" (I/8, 187). It is safe to assume that

Schelling intends this as a *modus tollens* against Eschenmayer's rejection of anthropomorphism and not a deduction of the validity of lifeless formalism, but he does not supply the missing premise showing why the rich structure of Christian presentation must be maintained and conversely why sobriety of expression would be toxic to religious belief.

The closest the *Freedom* essay comes to supplying this premise begins surprisingly with Schelling's statement that he is "convinced that reason is fully adequate to expose every possible error (in genuinely spiritual matters) and that the inquisitorial demeanor in the judgment of philosophical systems is entirely superfluous" (I/7, 412). Especially harmful to genuine philosophical insight is the sort of lazy relativism that makes history the judge of Christianity's truth by arguing that its demonstrable—by what means Schelling does not bother to question—success over paganism is proof enough of its rightness. This line of thinking fails because it cannot show how paganism remains as an independent ground that is superseded by Christianity and yet remains as a spiritual force indissoluble into the higher religion (I/7, 413). Thus while reason is sufficient to locate the flaws in any philosophical system, it can only find itself at home in a system whose content and form of presentation have developed from non-rational sources. "A system in which reason really recognized itself," he writes, "would have to unify all demands of the spirit as well as those of the heart and those of the moral feeling as well as those of the most rigorous understanding" (I/7, 413). By itself, the claim that a philosophical system should develop from nonreason into reason and thus that "inspiration in the genuine sense is the active principle of every productive and formative art or science" (I/7, 414) is nothing new for Schelling, since his *System of Transcendental Idealism* attempts to show how the unconscious develops (morally, aesthetically, etc.) into a reason fully conscious of itself.

Nor is Schelling departing from his earlier systems when he states, "Reason is not activity, like spirit, nor is it the absolute identity of both principles of cognition, but rather indifference; the measure and, so to speak, the general place of truth, the peaceful site in which primordial wisdom is received" (I/7, 415), for the entire *Presentation of My System of Philosophy* is built on the premise that reason is the indifferent site of truth's appearance. Rather, the claim that newly comes to the fore in the *Freedom* essay is that the ground-existence structure of personality gives us "an older revelation than any written one: nature. It contains prototypes that no human being has yet noted, whereas the written one received its fulfillment and interpretation long ago" (I/7, 415). But neither in the *Freedom* essay nor in the letter to Eschenmayer is the nature of these prototypes explored. While its quasi-dialogical form accomplishes a great deal, Schelling's correspondence with Eschenmayer leaves unanswered how philosophy can learn from nature to have a personality. How, he leaves us wondering, can reason find itself in a system in which it has been marginalized?

NOTES

Preface

1. For a brief appraisal of the contemporary Schelling reception, see my "Schelling's Contemporary Resurgence: The Dawn After the Night When All Cows Were Black," *Philosophy Compass* 6/9 (2011), 585–598.
2. For more on this problem, see my "The Reawakening of the Barbarian Principle," *The Barbarian Principle: Merleau-Ponty, Schelling, and the Question of Nature*, ed. Jason M. Wirth with Patrick Burke (Albany: State University of New York Press, 2013), chapter 1.
3. All citations from *Über das Verhältniß des Realen und Idealen in der Natur, oder Entwickelung der ersten Grundsätze der Naturphilosophie an den Principien der Schwere und des Lichts* (1806 version) are from a currently unpublished translation by my colleague and friend Iain Hamilton Grant.
4. *Anthropology, History, and Education*, eds. and trans. Robert B. Louden and Günter Zöller (Cambridge: Cambridge University Press, 2008), 438. The original reference can be found in the Akademie Ausgabe (*Werke*), 9: 442. See also John McCumber's discussion of this line in *On Philosophy: Notes from a Crisis* (Stanford: Stanford University Press, 2013), 15.
5. See, for example, Schelling's discussion in the *Historical-critical Introduction to the Philosophy of Mythology* of the "merely externally

humanlike races of South America" who apparently knew even less community than ants and honey bees and who "live like animals of the field" (II/1, 63) or his lament in the *Philosophie der Mythologie oder Darstellung der reinrationalen Philosophie* that, despite the horrors of slavery, such an encounter could at least have been an opportunity for the "salvation of lost souls" (II/1, 513). Although it is also true that for his time, Schelling often displays an extraordinarily generous interest in and appreciation for non-European cultures, he also had his limits. Perhaps we all do. It is important to be aware of and vigilant about these things in ourselves and in the texts that we read and to struggle against them, but it is also important to open up a text, to attempt to liberate its energies and creativity. In this sense it is my intent to think both with and through Schelling, to facilitate a coming to the fore of his powerful sense of philosophy.

Chapter 1. Extinction

1. Jean-Luc Nancy, *Being Singular Plural*, trans. Robert D. Richardson and Anne E. O'Byrne (Stanford, CA: Stanford University Press, 2000), 3. Henceforth BSP.
2. F.W.J. Schelling, *Clara*, trans. Fiona Steinkamp (Albany: State University of New York Press, 2002), 33. An earlier and much shorter version of this chapter appeared as "Mass Extinction: Schelling and Natural History," *Poligrafi: Journal for Interdisciplinary Study of Religion* (special issue: Natural History with special guest editor, David Michael Levin), no. 61–62, vol. 16 (2011), 43–63.
3. Except where noted, all translations of Schelling are my own responsibility. Citations follow the standard pagination, which follows the original edition of two divisions and fourteen volumes established after Schelling's death by his son, Karl. It lists the division, followed by the volume, followed by the page number. Hence, (I/1, 1) would read, division one, volume one, page one. It is preserved in Manfred Schröter's critical reorganization of this material. *Schellings Sämtliche Werke* (Stuttgart-Augsburg: J. G. Cotta, 1856–1861); *Schellings Werke: Nach der Originalausgabe in neuer Anordnung*, ed. Manfred Schröter (Munich: C. H. Beck, 1927). The first division is in ten books, and the second is in four books. Schröter numbers the volumes without division from I-XIV, so a citation that

might read SW XI would be for our present purposes II/1. My translations of *Die Weltalter* originally appeared in *The Ages of the World* (Albany: State University of New York Press, 2000). For citations of this work I also use the standard pagination and not the translation pagination, although the prior is embedded in the latter.

4. Io fui gia quel che voi siete e quell ch'io sono voi anco sarete.
5. See also Slavoj Žižek, *In Defense of Lost Causes* (London and New York: Verso, 2008), 442: "Humanity has nowhere to retreat to: not only is there no 'big Other' (self-contained symbolic order as the ultimate guarantee of Meaning); there is also no *Nature qua* balanced order of self-reproduction whose homeostasis is disturbed, nudged off its course, by unbalanced human interventions."
6. This is the now famous term coined by Paul Josef Crutzen and Eugene Stoermer in 2000 to describe the ascendant role of human manufacture in the prevalent geological character of the earth. The climate scientist William F. Ruddiman's provocative recent work, *Plows, Plagues, and Petroleum: How Humans Took Control of Climate* (Princeton: Princeton University Press, 2007), makes a case that the Anthropocene is not coterminous, as Cruzen and others have argued, with the Industrial Revolution, but rather with the interaction between human population numbers (their rise and fall, including the effect of pandemics on overall population numbers), cultural practices (starting with events like deforestation for agriculture, which increases as population numbers increase), and the climate. The implication of his argument is startling: there is not an independent, pure, healthy, ideal climate that would exist independent of industrialization (for the latter merely shifts the direction of the effect of the Anthropocene Age, namely, toward the widespread heating of the planet and the consequent alteration of the climate), but rather that whether the earth is heating (post-eighteenth century) or cooling (the enigmatic Little Ice Age that began in the mid-fourteenth century), the climate of the Anthropocene is inseparable from the widespread presence of human culture. That is nothing less than to say: already at a meteorological level it no longer makes sense to divorce "natural climate" from "human cultural activity." The two are rather parts of a more complex whole emerging out of the interplay of these two sets of trajectories. We are not *in* the climate as if it were our mere surroundings (environment). In some very real sense it makes more sense to say that we

are (of and toward) the climate. It is a feedback loop in which the ἄνθρωπος is a leading indicator species.

7. Maurice Merleau-Ponty, *Nature: Course Notes from the Collège de France*, ed. Dominique Séglard, trans. Robert Vallier (Evanston: Northwestern University Press, 2003), 47. For more on Merleau-Ponty's confrontation with Schelling's thinking, see *The Barbarian Principle: Merleau-Ponty, Schelling, and the Question of Nature*, ed. Jason M. Wirth with Patrick Burke (Albany: State University of New York, 2013).
8. Georges Bataille, *The Accursed Share*, vol. 1, trans. Robert Hurley (New York: Zone Books, 1991), 34.
9. Richard Leakey and Roger Lewin, *The Sixth Extinction: Patterns of Life and the Future of Humankind* (New York: Doubleday, 1995), 233. Henceforth SE.
10. Hence, Iain Hamilton Grant, in his *Philosophies of Nature after Schelling* (London and New York: Continuum, 2006), renders *unbedingt* literally as "unthinged." This is a watershed study of Schelling's *Naturphilosophie* in the English-speaking world. Henceforth PON.
11. For a discussion of Schelling's relationship to Spinoza, see my *The Conspiracy of Life: Meditations on Schelling and His Time* (Albany: State University of New York Press, 2003), chapters 2 and 3. For more on the problem of the image of thought, see chapter 3 of this present book.
12. On the problem of expressionism in Schelling, see my "Animalization: Schelling and the Problem of Expressivity," *Schelling Now: Contemporary Readings of Schelling*, ed. Jason M. Wirth (Bloomington: Indiana University Press, 2004), 84–98.
13. Jefferson bequeathed the fossilized jawbone of a mastodon to the University of Virginia.
14. Both the anecdote about Lewis and Clark and the Jefferson quotation are from Elizabeth Kolbert's essay, "The Sixth Extinction?" *New Yorker*, May 25, 2009, p. 53. Henceforth SE. In her eloquent and eye-opening essay, Kolbert also speaks with the Harvard paleontologist Andrew Knoll who reflects on the nature of the sixth extinction event: When the asteroid collided with the Yucatan Peninsula, "it was one terrible afternoon . . . But it was a short-term event, and then things started getting better. Today, it's not like you have a stress and the stress is relieved and recovery starts. It gets bad and then it keeps being bad, because the stress doesn't go away.

Because the stress is us" (63). See also Elizabeth Kolbert, *The Sixth Extinction: An Unnatural History* (New York: Holt, 2014).

15. See SE, 53–54.
16. SE, 54. Deleuze and Guattari argue in "10,000 B.C.: The Geology of Morals" that Cuvier and this kind of approach nonetheless is too tied to forms, not the unfolding of the earth: "There are irreducible axes, types, branches. There are resemblances between organs and analogies between forms, nothing more." *A Thousand Plateaus*, trans. Brian Massumi (Minneapolis: University of Minnesota, 1987), 46. For a more complicated account of Cuvier, see Grant, PON, 124–125.
17. Immanuel Kant, *Kritik der reinen Vernunft* (A edition 1781, B edition 1787) (Hamburg: Felix Meiner Verlag, 1976), 186. The citation appears in the A edition, 126–127.
18. Ibid., 462 (A 444/B 472–A 445/B 473).
19. *On the History of Modern Philosophy* (1827), trans. Andrew Bowie (Cambridge: Cambridge University Press, 1994), 173. Henceforth HMP with the German citation preceding the English. The German can be found at I/10, 177.
20. Immanuel Kant, *Kritik der Urteilskraft* (1790) (Hamburg: Felix Meiner Verlag, 1990), §22, 82.
21. Fyodor Dostoevsky, *The Brothers Karamazov*, trans. Richard Pevear and Larissa Volokhonsky (New York: FSG, 2002), 362.
22. David Wood, *The Step Back: Ethics and Politics after Deconstruction* (Albany: State University of New York Press, 2005), 7.
23. See also Axel Hutter, *Geschichtliche Vernunft: Die Weiterführung der Kantischen Vernunftkritik in der Spätphilosophie Schellings* (Frankfurt am Main: Suhrkamp, 1996), 266–270. "Hegel's philosophy is thereby itself its own kind of net that keeps the world imprisoned" (269–270).
24. *Urfassung Philosophie der Offenbarung*, two volumes, ed. Walter E. Ehrhardt (Hamburg: Felix Meiner Verlag, 1992), 659. Henceforth U. This is a much more complete and exciting version of the Philosophy of Revelation than the one that appeared in the edition prepared by Karl Schelling. It is derived from the 1831–1832 original version of the course and preserved in a carefully dictated hand-written manuscript, which the editor found in the university library of the Catholic university Eichstätt. It was an historic discovery.
25. Gary Snyder, *The Practice of the Wild* (Berkeley, CA: North Point Press, 1990), 185. Henceforth PW.

26. Gary Snyder, "Earth Day and the War Against the Imagination," *A Place in Space: Ethics, Aesthetics, and Watersheds* (Washington, DC: Counterpoint, 1995), 64.
27. Martin J.S. Rudwick, *Georges Cuvier, Fossil Bones, and Geological Catastrophes: New Translations & Interpretations of the Primary Texts* (Chicago and London: University of Chicago Press, 1997), 261–262. Henceforth GC.
28. John Sallis, *Force of Imagination: The Sense of the Elemental* (Bloomington and Indianapolis: Indiana University Press, 2000. Henceforth FI.
29. See Dorinda Outram, *Georges Cuvier: Vocation, Science, and Authority in Post-Revolutionary France* (Manchester: Manchester University Press, 1984), 135–138. Henceforth GCV.
30. On the problem of logic and the imagination, see Sallis' important study, *Logic of Imagination: The Expanse of the Elemental* (Bloomington and Indianapolis: Indiana University Press, 2012). "The consequences for logic are evident. What is required is a change of focus, a shift, as it were, of the center of gravity: the focus will no longer be on the concept, nor consequently on the proposition, judgment, syllogism, or more generalized inference forms. Rather, the logic of the concept is to give way to a logic focused on the form that primary determination assumes when it is no longer prescribed for the sensible from without (for instance, thorough categories) but is reinstalled in the sensible" (22).
31. There are so many editions of *Moby-Dick*, so I will cite the novel by chapter number and title. I use the critical edition established in the Northwestern-Newberry Edition of the Writings of Herman Melville, vol. 6, eds. Harrison Hayford, Hershel Parker, and G. Thomas Tanselle (Evanston and Chicago: Northwestern University Press and the Newberry Library, 1988).
32. This is the ninth of the 1805 *Aphorismen zur Einleitung in die Naturphilosophie*.

Chapter 2. Solitude of God

1. See Bruce Matthews: "His rather heretical strategy to move ahead is to cast this estrangement in the subject-object terms of our relation to nature: the *Entzweiung* is between the illusive asymmetrical

relationship we have posited to exist between the human subject and the object of nature. This is the context in which his critique of modern philosophy since Descartes occurs: that making the subject absolute will lead to the *annihilation* of nature. The *estrangement* requiring resolution, however, is not *within* the human self per se, but rather between the human subject and the object of nature. In the closing paragraph of his *Freedom* essay Schelling expresses this eloquently, connecting the 'possibility of immediate knowledge' with access to a 'revelation' of truth more originary than that of the religious historical faith. The revelation of which he speaks is to be sought in the forces of nature: . . . 'The time of merely historical faith is past as soon as the possibility of immediate knowledge is given. We have an earlier revelation than any written one—nature . . . If the understanding of that unwritten revelation were inaugurated, the only true system of religion and science would appear. . . .'" Bruce Matthews, *Schelling's Organic Form of Philosophy: Life as the Schema of Freedom* (Albany: State University of New York Press, 2011), 32.

2. "Elohim called unto him out of the midst of the bush . . ."
3. *Zen Sourcebook: Traditional Documents from China, Korea, and Japan*, ed. Stephen Addiss, with Stanley Lombardo and Judith Roitman (Indianapolis and Cambridge: Hackett, 2008), 4. Henceforth ZS.
4. As is well known, this term stems from Duns Scotus' *haecceitas*, the irreducible and nonsubsumable "thisness" of each being.
5. Georges Bataille, *Theory of Religion*, trans. Robert Hurley (New York: Zone Books, 1989), 53.
6. See my "The Language of Natural Silence: Schelling and the Poetic Word After Kant," *The Linguistic Dimension of Kant's Thought: Historical and Critical Essays*, eds. Frank Schalow and Richard Velkley (Evanston: Northwestern University Press, 2014), 237–262.
7. Es muβ *vor* allem Grund und vor allem Existierenden, also überhaupt vor aller Dualität, ein Wesen sein; wie können wir es anders nennen als den Urgrund oder vielmehr *Ungrund*?
8. Gilles Deleuze and Félix Guattari, *What is Philosophy?*, trans. Hugh Tomlinson and Graham Burchell (New York: Columbia University Press, 1994), 43. Henceforth WP.
9. Gianni Vattimo, *After Christianity*, trans. Luca D'Isanto (New York: Columbia University Press, 2002), 30. Henceforth AC.
10. Michel Foucault, *Politics, Philosophy, Culture: Interviews and Other*

Writings, 1977–1984, ed. Lawrence D. Kritzman (New York: Routledge, 1988), 328.

11. Das Vergangene wird gewußt, das Gegenwärtige wird erkannt, das Zukünftige wird geahndet. Das Gewußte wird erzählt, das Erkannte wird dargestellt, das Geahndete wird geweissagt.
12. Hermann Broch, *The Death of Virgil* (1945), trans. Jean Starr Untermeyer (New York: Vintage International, 1995), 178. Henceforth DV.
13. See Slavoj Žižek, *The Indivisible Remainder: An Essay on Schelling and Related Matters* (London and New York: Verso, 1996). He notes, in a Lacanian idiom, that for Schelling there "must be a signifier (a 'something') which stands for 'nothing,' a signifying element whose very presence stands for the absence of meaning (or, rather, for absence tout court). This 'nothing,' of course, is the subject itself, the subject qua $, the empty set, the void which emerges as the result of the contraction in the form of expansion: when I contract myself outside myself, I deprive myself of my substantial content" (43–44).
14. Lore Hühn, *Kierkegaard und der deutsche Idealismus: Konstellationen des Übergangs* (Tübingen: Mohr Siebeck, 2009), 236–239.
15. Friedrich Nietzsche, *Thus Spoke Zarathustra*, trans. Graham Parkes (Oxford: Oxford University Press, 2005), 239. Zarathustra is admonishing his shadow.
16. Sri Aurobindo Ghose, *Essays on the Gītā* (Pondicherry: Sri Aurobindo Ashram, 1970), 42.
17. See, for example, Bernard, Freydberg's thoughtful study, *Schelling's Dialogical* Freedom essay: *Provocative Philosophy Then and Now* (Albany: State University of New York Press, 2008). His point about the mythic quality of the essay, especially in the Platonic sense of myth as a likely story, is well taken. It is also important to consider the extent to which the *Freedom* essay is not just a "likely account" or a story about human freedom any more than it is theory of human freedom. It is the revelation of freedom, a *tautegorical* revelation of the Godhead. The Godhead reveals itself in the consuming fire and Great Death of nature. In mythology, the gods may come as themselves, but the revelatory character of appearance is not itself revealed. Nonetheless, revelation needs its own mythology. For more on this and related issues, see my "Schelling and the Future of God," *Analecta Hermeneutica*, vol. 5.

18. *Initia Philosophiæ Universæ* (1820–1821), ed. Horst Fuhrmans (Bonn: H. Bouvier, 1969), 18. Henceforth IPU. This is an early critical edition of the *Erlanger Vorlesungen*. These lectures formed the basis of Schelling's essay, *On the Nature of Philosophy as Science* (1821). A new and more complete version is now in the works.
19. *Philosophie der Offenbarung* (1841–1842), ed. Manfred Frank (Frankfurt am Main, 1977), 316. Henceforth PO.
20. Schelling is referring to the Book of Revelation 21:2.
21. This is from the *Historical-Critical Introduction to the Philosophy of Mythology* (1842), trans. Mason Richey and Markus Zisselsberger (Albany: State University of New York Press, 2007), 174. Henceforth HCI.

Chapter 3. Image of Thought

1. Ruth Stone, *In the Next Galaxy* (Port Townsend, Washington: Copper Canyon Press, 2002), 15.
2. Gilles Deleuze, *Difference and Repetition*, trans. Paul Patton (New York: Columbia University Press, 1994). *Différence et répétition* (Paris: Presses universitaires de France, 1968). Henceforth DR, with the English citation followed by the French citation.
3. "Thought is like the Vampire; it has no image, either to constitute a model of or to copy. In the smooth spaces of Zen, the arrow does not go from one point to another but it is taken up at any point, to be sent to any other point, and tends to permute with the archer and the target." Deleuze and Guattari, *A Thousand Plateaus*, trans. Brian Massumi (Minneapolis: University of Minnesota Press, 1987), 377.
4. For a provocative series of confrontations with this problem (the remains of God language of any sort whatsoever) in reference to the figures germane to our essay, see Anthony Paul Smith and Daniel Whistler, eds., *After the Postsecular and the Postmodern: New Essays in Continental Philosophy of Religion*. (Newcastle: Cambridge Scholars' Press, 2010. See especially Whistler's essay, "Language after Philosophy of Nature: Schelling's Geology of Divine Names" (335–359).
5. Gilles Deleuze, *Nietzsche and Philosophy*, trans. Hugh Tomlinson (New York: Columbia University Press, 1983), 4. *Nietzsche et la*

philosophie (Paris: Presses universitaires de France, 1962), 4. Henceforth NP, with the English citation followed by the French citation.

6. *Nietzsche Werke: Kritische Gesamtausgabe*, vol. 6:1, eds. Giorgio Colli and Mazzino Montinari (Berlin: Walter de Gruyter, 1968), 226. All translations from German language materials are my own responsibility, unless otherwise noted.
7. *Nietzsche Werke: Kritische Gesamtausgabe*, vol. 6:3, eds. Giorgio Colli and Mazzino Montinari (Berlin: Walter de Gruyter, 1969), 183.
8. Deleuze goes on to ask, "How else can one write but of those things which one doesn't know, or knows badly?" He then suggests that "perhaps writing has a relation to silence altogether more threatening than that which it is supposed to entertain with death" (DR, xxi/4). This sensibility guides this present chapter.
9. *On the History of Modern Philosophy* (1827), trans. Andrew Bowie (Cambridge: Cambridge University Press, 1994), 160.
10. Deleuze makes the same of point about Hegel: "Difference is the ground, but only the ground for the demonstration of the identical. Hegel's circle is not the eternal return, only the infinite circulation of the identical by means of negativity. Hegel's innovation is the final and most powerful homage rendered to the old principle" (DR, 50/71). This circle extends representation, radically and ingeniously, to unprecedented powers of conservation, to the inclusion of every other center into its infinite pantheon, into an infinite archive in which all difference will be a reprise or recapitulation of the same circle of thinking. "There is indeed a dialectical circle, but this infinite circle has everywhere only a single center; it retains within itself all other circles, all the other momentary centers. The reprises or repetitions of the dialectic express only the conservation of the whole [*seulement la conservation du tout*], all the forms and the moments, in a gigantic Memory. Infinite representation is a memory which conserves. In this case, repetition is no more than a conservatory, a power of memory itself" (DR, 53–54/76).
11. Hegel, along with Leibniz, comprise the extremes of representation, what Deleuze calls "orgiastic representation" (DR, 42/61)—the tumult of infinite combinatorial possibilities of representation, which, in rendering representation orgiastic, preserve representation as such. Leibniz and Hegel do this on opposite sides of the spectrum. For Leibniz is a question of orgiastic representation as

the problem of the infinitely small (monads) and for Hegel it is the problem of the infinitely large.

12. *Die Weltalter in den Urfassungen von 1811 und 1813 (Nachlaßband)*, ed. Manfred Schröter (Munich: Beck, 1946), 43. Henceforth WA.
13. Translations of this work follow my *The Ages of the World* (1815 draft) (Albany: State University of New York Press, 2000).
14. Both Schelling and Deleuze embody vast philosophical enterprises. In the interest of a wieldy confrontation between the two, I generally stick to Deleuze's work from the 1960s, especially the works of 1968, and I chiefly focus on Schelling's middle period, especially the *Freedom* essay and the drafts of *Die Weltalter*.
15. This collection includes excerpts from the 1800 *System*, *Ideas for a Philosophy of Nature*, *Von der Weltseele*, *Erster Entwurf*, *Darstellung des Philosophischen Empiricismus*, and the *Freedom* essay.
16. Despite Deleuze's misgivings about the inability of the image of the abyss to sustain difference, in *The Logic of Sense* he nonetheless admires Schelling's (as well as Böhme's, Schopenhauer's, and the early Nietzsche's) "moments in which philosophy makes the Abyss [*Sans-fond*] speak." *The Logic of Sense*, trans. Mark Lester with Charles Stivale (New York: Columbia University Press, 1990), 106–107. Deleuze will again argue that behind the speaker, however, "*you will discern* nothing" (107), which, in the end, evacuates the singularities of difference. I will question the aptness of this indictment of Schelling later on.
17. From the *Nachlaß*:

    > Nihilism. It is *ambiguous*:
    >
    > Nihilism as symptom of the enhanced power of the spirit [*Zeichen der gesteigerten Macht des Geistes*]: active nihilism.
    >
    > Nihilism as the decline and regression of the power of the spirit: passive nihilism.
    >
    > Friedrich Nietzsche, *Werke in Drei Bänden*, ed. Karl Schlechta (Munich: Carl Hanser Verlag, 1956), 557.

18. "All beginning would lie in lack [*Mangel*], and the deepest potency on which all is hinged is what does not have being [*das Nichtseyende*] and this is the hunger for Being" (II/1, 294).

19. Gilles Deleuze, *Expressionism in Philosophy: Spinoza*, trans. Martin Joughin (New York: Zone Books, 1990), 118. Henceforth EPS.
20. For an account of this, see my *The Conspiracy of Life: Meditations on Schelling and His Time* (Albany: State University of New York Press, 2003), chapters 2 and 3.
21. See Robert F. Brown's *The Later Philosophy of Schelling: The Influence of Böhme on the Works of 1809–1815* (Lewisburg: Bucknell University Press, 1977).
22. Ernst Benz, *Schellings theologische Geistesahnen* (Wiesbaden: Akademie der Wissenschaften und der Literatur, in Kommission bei Franz Steiner Verlag, 1955).
23. Bruce Matthews, *Schelling's Organic Form of Philosophy: Life as the Schema of Freedom* (Albany: State University of New York Press, 2011), 52.
24. Alberto Toscano, "Philosophy and the Experience of Construction," *The New Schelling*, eds. Judith Norman and Alistair Welchman (London and New York, 2004), 123. One could also hear this resonate in Matthews' title and its evocation of the organic form of philosophy.
25. We can here cite Christian Kerslake: "In the section on Spirit in the *Phenomenology of Spirit*, 'we' are told about our prehistory, which is 'recollected [*erinnert*]' and brought into consciousness; after we have read the *Phenomenology of Spirit*, we should be able give full justification for our beliefs, because there is only one history of Spirit. However, there is something destabilizing and uncanny about the entrance to Schelling's late system." *Immanence and the Vertigo of Philosophy: From Kant to Deleuze* (Edinburgh: Edinburgh University Press, 2009), 87.
26. Deleuze: "Recognition is a sign of the celebration of monstrous nuptials, in which thought 'rediscovers' the State, rediscovers 'the Church' and rediscovers all the current values that it subtly presented in the pure form of an eternally blessed unspecified eternal object" (DR, 136/177).
27. Deleuze: "Schelling said that depth is not added from without to length and breadth, but remains buried, like the sublime principle of the *differend* which creates them" (DR, 230/296).
28. See also in this respect the important 1806 study (an addendum or *Zugabe* to the 1798 *Weltseele* text), "On the Relation of the Ideal and the Real in Nature, or Development of the First Principles of

Naturephilosophy from the Principles of Gravity and Light [*Über das Verhältniß des Realen und Idealen in der Natur, oder Entwicklung der ersten Grundsätze der Naturphilosophie an den Principien der Schwere und des Lichts*]." References to this seminal text borrow from Iain Hamilton Grant's invaluable but as yet unpublished translation with occasional alterations in the interest of maintaining stylistic continuity with my own translations.

29. Jean-François Lyotard, *The Differend: Phrases in Dispute*, trans. Georges van den Abbeele (Minneapolis: University of Minnesota Press, 1989), 97. Henceforth D.
30. ". . . in den heiligen Sabbath der Natur einführen, in die Vernunft, wo sie, ruhend über ihren vergänglichen Werken, sich selbst als sich selbst erkennt und deutet. Denn in dem Maß, als wir selbst in uns verstummen, redet sie zu uns."
31. Lore Hühn's *Kierkegaard und der deutsche Idealismus: Konstellationen des Übergangs* contributes significantly to the problem of passage (*Übergang*) that is key to our current chapter.
32. One might say that thinking does not complete itself because it is already complete, but by that we are saying that completeness does not derive from something striving to fill a lack, to seal up an empty space. For Schelling, the infinite lack (*Mangel*) of being is its fullness, its enduring love of its own becomings. The infinite depth of gravity belongs to the consummation of nature, not its deficit.
33. Todd May, *Gilles Deleuze: An Introduction* (Cambridge: Cambridge University Press, 2005), 20.
34. Thomas P. Kasulis, *Intimacy or Integrity: Philosophy and Cultural Difference* (Honolulu: University of Hawaii Press, 2002), 53.
35. "Hören Sie denn nichts? hören Sie denn nicht die entsetzliche Stimme, die um den ganzen Horizont schreit und die man gewöhnlich die Stille heißt? Seit ich in dem stillen Tal bin, hör ich's immer, es läßt mich nicht schlafen." Georg Büchner, *Lenz* (1879), in *Werke und Briefe*, vol, 1, ed. Fritz Bergemann (Frankfurt am Main: Insel Verlag, 1982), 110.
36. Milan Kundera, *Farewell Waltz*, trans. Aaron Asher (New York: HarperCollins, 1998), 273. Henceforth FW.
37. Edgar Allen Poe, "The Imp of the Perverse," *Tales of Edgar Allen Poe* (New York: Random House, 1944), 441.
38. *Stuttgarter Privatvorlesungen* (1810), unedited version, ed. Miklos Vetö (Turin: Bottega d'Erasmo, 1973), 152.

39. *Nietzsche Werke: Kritische Gesamtausgabe*, vol. III:1, eds. Giorgio Colli and Mazzino Montinari (Berlin: Walter de Gruyter, 1972), 333.
40. "A distribution is in conformity with good sense when it tends to banish difference from the distributed" (DR, 224/289).
41. See also, "Philosophy becomes Genetic: The Physics of the World Soul," *The New Schelling*, eds. Judith Norman and Alistair Welchman (London and New York, 2004), 128–150.
42. Deleuze: "Expecting repetition from the law of nature is the 'Stoic' error" (DR, 3/10).
43. As Merleau-Ponty articulates this in his return to Schelling and the question of nature in his Collège de France course, "Nature is always new in each perception, but it is never without a past. Nature is something which goes on, which is never grasped at its beginning, though it appears always new to us." *La nature: Notes, cours du Collège de France*, ed. Dominique Séglard (Paris: Seuil, 1995), 160.
44. Deleuze: "A single and same voice for the whole thousand-voiced multiple, a single and same Ocean for all the drops, a single clamor of Being for all beings [*une seule clameur de l'Être pour tous les étants*]: on condition that each being, each drop and each voice has reached the state of excess—in other words, the difference which displaces and disguises them and, in turning upon its mobile cusp, causes them to return" (DR, 304/388–389).
45. Henri Bergson, *Matter and Memory* (1896), trans. N.M. Paul and W.S. Palmer (New York: Zone Books, 1988), 191. Henceforth MM.
46. Concerning this, I must respectfully disagree with Christopher Groves who, in defending Deleuze against Schelling, rejects what he deems to be Schelling's "essentially foundationalist project" (27) and his "dogmatic, foundationalist tendency" (44). "Schellingian reason undermines its own claims through its rigorously foundationalist method" (43). I regard Groves' approach as another variant on Jaspers' *Größe und Verhängnis* strategy, namely, to treat Schelling as unable to sustain the force of his most radical insight into the disequilibrium of the being of becoming. Hence, despite the *Verhängnis* of Schelling's identity philosophy, one should opt for the subterranean greatness of Schelling's "minor" tendency to pursue this more unsettling dimension of his thinking. "Further, this renunciation of the universal and representative as the goal of thought can only be pursued by affirming the import of

the subversive, 'minor' tendency of Schelling's work, that is, that the Absolute must be thought as incommensurable with itself" (44). Christopher Groves, "Ecstasy of Reason, Crisis of Reason: Schelling and Absolute Difference," *Pli* 8 (1999), 25–45.

47. Although the extent to which Schelling dramatizes disparity varies throughout his thinking, it is always central. Homogenous time and space are not the transcendental conditions per se, but rather the incommensurablity of eternity and time, chaos and place, is the condition for the nonlinear recapitulation of nature and the self. Identity and indifference for Schelling serve reject the transcendence of an ontotheological God in favor for the divine depths of the earth, depths that can neither be conflated with the earth nor detached from it. Deleuze himself remarks in passing that Schelling's *Ungrund*, like the Dionysus of the early Nietzsche and the thought of Schopenhauer, "cannot sustain difference." "Since groundlessness lacks both individuality and singularity, it is therefore necessarily represented as devoid of any difference." Groundlessness is represented "as a completely undifferentiated abyss, a universal lack of difference, an indifferent black nothingness" (DR, 276/354). I do not find Deleuze's indictment of Schelling on this issue consonant with Deleuze's own defense of Schelling from Hegel's denigration of the implicitly Schellingian "night when all cows are black." I also do not think that Deleuze, at least in this passage, appreciates what I am here calling the solitude of God. In what manner can *der ewige Anfang* be confused with "an indifferent black nothingness"? Is this the way to read *das Seinskönnende* and the μὴ ὄν? Was this aspect of groundlessness not precisely what the positive philosophy moved to deny? In the *Freedom* essay, Schelling claims that "Hence we have demonstrated the determinate point of the system where the concept of indifference is surely the only one possible for the absolute. If this concept is however taken as a generality, the whole is distorted and it would then follow that this system sublates the personality of the supreme *Wesen* [*die Persönalität des höchsten Wesens aufhebe*]. In the *Ungrund* or indifference there is no personhood [*Persönlichkeit*], but is the point of departure then the whole" (I/7, 412)? I take the singularity of personality, *God's solitude*, to run counter to the impersonality of treating the problem of the *Ungrund* as simply an abstraction or a generality about absolute nothingness.

48. As we have seen, what is characteristically *modern* in modern philosophy is pernicious: "that nature is not present to it, and that it lacks a living ground [*daß die Natur für sich nicht vorhanden ist, und daß es ihr am lebendigen Grunde fehlt*]" (I/7, 361).
49. "The eternal return is the same *of* the different, the one *of* the multiple, the resemblant *of* the dissimilar [*le meme* du *different, l'un* du *multiple, le ressemblant* du *dissemblable*]" (DR, 126/165).
50. "*Revenir, l'être de ce qui devient*" (NP, 48/55).
51. At *Monadology* 47, Leibniz argues that monads are a "constant fulguration."

Chapter 4. Stupidity

1. Avital Ronell, *Stupidity* (Urbana and Chicago: University of Illinois Press, 2002), 37. An earlier and shorter version of this chapter appeared in *The Weight of Violence: Religion, Language, Politics*, eds. Saitya Brata Das and Soumyabrata Choudhury (Oxford India, 2015).
2. Rustam Singh, *"Weeping" and Other Essays on Being and Writing* (Jaipur, India: Pratilipi Books, 2011), 164.
3. In *Conversations with Nietzsche: A Life in the Words of His Contemporaries*, ed. Sander L. Gilman, trans. David J. Parent (New York and Oxford: Oxford University Press, 1987), 230. Henceforth CN.
4. Robert Musil, "On Stupidity," *Precision and Soul*, trans. Burton Pike and David S. Luft (Chicago and London: University of Chicago Press, 1990). Henceforth OS.
5. Gilles Deleuze, *Nietzsche and Philosophy*, trans. Hugh Tomlinson (New York: Columbia University Press, 1983), 105. *Nietzsche et la philosophie* (Paris: Presses universitaires de France, 1962), 120. Henceforth NP with the English citation followed by the French citation.
6. Robert Musil, *The Man Without Qualities*, vol. 1, trans. Sophie Wilkins (New York: Vintage, 1996). Henceforth MWQ.
7. Gustave Flaubert, *Bouvard and Pécuchet*, trans. T.W. Earp and G.W. Sonier (New York: New Directions, 1954). Henceforth BP.
8. Jorge Luis Borges, "A Defense of *Bouvard et Pécuchet*" (1954), *Selected Non-Fictions*, ed. Eliot Weinberger (New York: Viking, 1999), 386. Henceforth DBP.

9. Gustave Flaubert, *The Dictionary of Accepted Ideas*, trans. and ed. Jacques Barzun, included as a supplement to BP, second part, 47. Henceforth DAI.
10. Jacques Derrida, *The Beast & the Sovereign*, vol. 1, trans. Geoffrey Bennington (Chicago and London: University of Chicago Press, 2009), 158. Henceforth BS.
11. Der Mensch bekommt die Bedingung nie in seine Gewalt, ob er gleich im Bösen danach strebt . . . Daher der Schleier der Schwermut, der über die ganze Natur ausgebreitet ist, die tiefe unzerstörliche Melancholie alles Lebens.
12. *Initia Philosophiæ Universæ* (1820–1821), ed. Horst Fuhrmans (Bonn: H. Bouvier, 1969), 6–7.
13. See also WA, 41. *Angst* is the *Empfindung* that corresponds to the conflict of directions in Being. Amid this strife, "the cision comes forth and brings the forces to ever greater severance so that the contracting force, so to speak, trembles for its existence." Or AW, 246: "Hence, since the first potency unites within it conflicting forces, of which one always craves the outside and of which the other is always inwardly restrained, its life is a life of loathing [*Widerwärtigkeit*] and anxiety since it does not know whether to turn inward or outward and in this fashion falls prey to an arbitrary, revolving motion."
14. See Buchheim, 145. Buchheim also notes that a discourse of anxiety is also to be found in Böhme (especially *De signatura rerum*) as well as Friedrich Christoph Ötinger (1702–1782).
15. Michel Foucalt, "Theatrum Philosophicum," trans. Donald F. Brouchard and Sherry Simon, *Aesthetics, Method, and Epistemology: Essential Works of Foucault*, vol. 2, ed. James D. Faubion (New York: New Press, 1998), 361. Henceforth TP.
16. Julián Ferreyra, "Flaubert's *Bêtise* as a Way into Deleuzian Ideas." Unpublished manuscript presented at Harvard University, February 16, 2012.
17. Ibid.
18. Alfred North Whitehead, *Science and the Modern World* (1926) (Cambridge: Cambridge University Press, 2011), 245. Henceforth SMW.
19. For more on the problem of evil, see Wirth, *The Conspiracy of Life: Meditations on Schelling and His Time* (Albany: State University of New York Press, 2003), chapter 6.

20. This was the 1797 *System* fragment's infamous political anthem: "*Über den Staat hinaus*! Through the state beyond the state!—For each state must treat free humans as a mechanical cog" (*Materialien zu Schellings philosophischen Anfängen*, 110).
21. Fyodor Dostoevsky, *Notes from Underground*, trans. Richard Pevear and Larissa Volokhonsky (New York: Alfred A. Knopf, 1993), 24. Henceforth NU.
22. See the invaluable work, *The Philosophical Rupture Between Fichte and Schelling*, ed. and trans. Michael Vater and David W. Wood (Albany: State University of New York Press, 2012).
23. "Immanenz und Transcendenz völlig und gleich leere Worte sind, da sie eben selbst diesen Gegensatz aufhebt, und in ihr alles zusammenstießt zu Einer Gott-erfüllten Welt." This is from the 1806 *Weltseele* addendum on the relationship of gravity and light.
24. Richard Dawkins, *The God Delusion* (New York: Houghton Mifflin Harcourt, 2006), 360.
25. Ibid.
26. *Philosophische Einleitung in die Philosophie der Mythologie oder Darstellung der reinrationalen Philosophie*, delivered in Berlin between 1847 and 1852.
27. This distinction can be found in a work that Fénelon authored with Madame Jeanne Guyon entitled *Spiritual Progress: Instructions in the Divine Life of the Soul*. See book 10 (On Self-Abandonment): "If you would fully comprehend the meaning of self-abandonment, recall the interior difficulty which you felt, and which you very naturally testified when I directed you always to count as nothing this self which is so dear to us. To abandon one's self is to count one's self as nought; and he who has perceived the difficulty of doing it, has already learned what that renunciation is, which so revolts our nature. Since you have felt the blow, it is evident that it has fallen upon the sore spot in your heart; let the all-powerful hand of God work in you as he well knows how, to tear you from yourself. The origin of our trouble is, that we love ourselves with a blind passion that amounts to idolatry. If we love anything beyond, it is only for our own sakes. We must be undeceived respecting all those generous friendships, in which it appears as though we so far forgot ourselves as to think only of the interests of our friend. If the motive of our friendship be not low and gross, it is nevertheless still selfish; and the more delicate, the more concealed, and the more proper in

the eyes of the world it is, the more dangerous does it become, and the more likely to poison us by feeding our self-love." Edited and translated by James W. Metcalf (New York: Dodd, 1853), 60–62.

28. I return to this issue in chapter 6 in my discussion of *das Sinnbild.*
29. In the 1797 *Ideen zu einer Philosophie der Natur*, we find a remarkable line: "The ancients and after them the moderns quite significantly designated the real world as *natura rerum* or the birth *of things* [*die Geburt der Dinge*]; for it is in the real part that the eternal things or the ideas come into existence" (I/2, 187–188). Here Schelling reads Lucretius' *De rerum natura* not as the essence or nature of things, but as the natality that is already at the etymological heart of *natura* itself (from *natus* "born," past participle of *nasci*, "to be born"). The question of nature is the question of its *ewiger Anfang*, its ceaseless yet unprethinkable creativity, *the life of the imagination.*
30. See the 1815 *Ages of the World*: "For in entering its appropriate locus, it knows itself only as a potency and knows a higher potency above it. It becomes Being in relationship to this higher potency so that this higher potency can be active in it as in its own matter or immediate element. While it always remains, with respect to itself, what it is, namely, an eternal Yes that holds the negating force with it and conceals it, it is not a contradiction if that higher potency (A^3) liberates the negating force in the second potency and unfolds it with level-headedness and intention into another world. For its nature is just to be the originary affirmative principle that confines the dark primordial force" (I/8, 250). Also: "And, in the end, and with level-headedness, and without insult to the force which contains it and, so to speak, nourishes it, the soul is pleased by subordinating all forces through a piecemeal progression and thereby displaying its own mother, in whom the soul was first conceived and fostered, as a universally animated essence" (I/8, 278). This is the enchoric gesture that Schelling names the *religious.*
31. Ὁ λόγος γὰρ ὁ τοῦ σταυροῦ τοῖς μὲν ἀπολλυμένοις μωρία ἐστίν, τοῖς δὲ σῳζομένοις ἡμῖν δύναμις θεοῦ ἐστιν.
32. Stanislas Breton, *The Word and the Cross*, trans. Jacquelyn Porter (New York: Fordham University Press, 2002), 8.
33. *The Analects of Confucius: A Philosophical Translation*, trans. Roger T. Ames and Henry Rosemont Jr. (New York: Random House, 2010), 101.

Chapter 5. Plasticity

1. Barnett Newman, "The New Sense of Fate," *Barnett Newman: Selected Writings and Interviews*, ed. John P. O'Neill (Berkeley: University of California Press, 1990), 168. An earlier and shorter version of this chapter appeared in Italy as "Saturated Plasticity: Art and Nature," in *SpazioFilosofico*, issue 6: *Saturazione* [*Saturation*] (2012), 399–410.
2. Gary Snyder, "Preface" (2010), PW, ix.
3. I also here invoke Foucault's use of this image in his famous Nietzsche essay, where he refers to Nietzsche's eschewal of "solid identities" and his embrace of the "great carnival of time where masks are constantly reappearing." Michel Foucault, "Nietzsche, Genealogy, History," *The Foucault Reader*, ed. Paul Rabinow (New York: Pantheon, 1984), 94.
4. In such a call, Schelling has first and foremost the "fatherland" in view (I/7, 328). The question of specifically German earth and soil is quite complex. For a discussion of this problem, see Devin Zane Shaw, *Freedom and Nature in Schelling's Philosophy of Art* (New York: Continuum, 2010). Here Shaw presents what he considers both the bad news and the good news regarding the political ramifications of Schelling's philosophy of art. Schelling had hoped that the philosophy of art could unite German-speaking world in a new mythology, beyond its sectarian tendencies and the corruption of a culture that denigrates all value into use-value. Shaw finds this ambition dangerous: "A mythologization of politics is closer to what Marx would call a mystification: an idealization of social relationships . . . it places *Bildung* or cultivation of peoples or publics over direct or democratic political engagement" (116). Schelling would eventually abandon the project of a new mythology, and not a moment too soon. "That Schelling turns away from the mythologization of politics when he changes his focus to a universal history of religion—just at the time when German politics becomes increasingly nationalistic—appears well advised in retrospect" (117). As much as Schelling dreamed of the unifying force of a new mythology, there is also an undeniable radicality to his insight into the utopian promise of art. In an increasingly globalized and totally administered world, the eruption of freedom testified by the work of art continues to be an inspiring political resource. "On the

other hand, the revolutionary and utopian idea of art reemerged in the avant-garde of the 20th century and is still the focus of contemporary debates on the relationship between politics and art" (117). Schelling wrote no treatises on the political, and his passing comments on matters political continue to be a matter of earnest debate. Shaw for his part takes a strong stand: "If my concluding critique . . . is sharp or even polemical, it is only because I think the potential of the revolutionary sequence has yet to be exhausted" (7). I would simply add to Shaw's well-taken point that Schelling in the Munich essay does warn that "without great general enthusiasm, there are only sects" (I/7, 327), that is, without an awakening to the source of art, art quickly collapses into the nationalization of particular forms of art. One need only think of Hitler's attack on "*entartete Kunst*," art that betrays its form, and his promotion of "German" art, which, as hollow, politicized forms of art, was nothing but kitsch. In radical evil, ground and earth are covered with *Boden*. German soil is understood in its national and cultural form, detached from its living earth. On the question of earth, see also Rodolphe Gasché, *Geophilosophy: On Gilles Deleuze and Félix Guattari's* What is Philosophy? (Evanston: Northwestern University Press, 2014).

5. F.W.J. Schelling, *Ideas for a Philosophy of Nature*, trans. Errol E. Harris and Peter Heath (Cambridge: Cambridge University Press, 1988), 264–265.
6. Gilles Deleuze, *Francis Bacon: The Logic of Sensation*, trans. Daniel W. Smith (Minneapolis: University of Minnesota Press, 2003), 89.
7. For a concise account of this, see John R. Betz, "Reading 'Sibylline Leaves': J. G. Hamann in the History of Ideas," *Hamann and the Tradition*, ed. Lisa Marie Anderson (Evanston, Northwestern University Press, 2012), 5–32; see especially 14–19.
8. For reasons that will become clear later in the chapter, I leave the word *Wesen* untranslated.
9. Aldo Leopold, *A Sand County Almanac and Sketches Here and There* (1949) (Oxford: Oxford University Press, 1968). Henceforth SCA.
10. Holmes Rolston III, "Challenges in Environmental Ethics," *Environmental Philosophy: From Animal Rights to Radical Ecology*, fourth edition, eds. Michael E. Zimmerman, J. Baird Callicott, Karen J. Warren, Irene J. Klaver, John Clark (Upper Saddle River, NJ: Pearson, 2005), 82.

11. In the third or 1815 draft of *Die Weltalter*, Schelling makes this distinction about the Godhead: "But the good is its Being *per se*. It is essentially good and not so much something good as the Good itself" (I/8, 237). See also chapter one of my *Conspiracy of Life: Meditations on Schelling and His Time*.
12. This is from Dōgen's 1243 fascicle, *Sansuikyō* [*Mountains and Waters Sutra*], from the recent two-volume edition of the *Shōbōgenzō*, ed. Kazuaki Tanahashi (Boston and London: Shambhala, 2010), 160. Henceforth S.
13. Schelling, *System des transzendentalen Idealismus* (1800), eds. Horst D. Brandt and Peter Müller (Hamburg: Felix Meiner, 1992). Henceforth ST.
14. For more on the problem of μίμησις in Schelling's philosophy, see my "Schelling and the Force of Nature," *Interrogating the Tradition*, eds. John Sallis and Charles Scott (Albany: State University of New York Press, 2000), 255–274.
15. Wassily Kandinsky, *Über das Geistige in der Kunst* (1911) (Bern: Benteli Verlag, 1952), 21.
16. Georges Bataille, *The Accursed Share*, vol. 2 and 3, trans. Robert Hurley (New York: Zone Books, 1993), 204–206.
17. Catherine Malabou, *Plasticity at the Dusk of Writing: Dialectic, Destruction, Deconstruction*, trans. Carolyn Shread (New York: Columbia University Press, 2010). Henceforth PDW.
18. This photograph of a Kwakiutl (Kwak'wala) transformational mask by John Livingston (1951–), adopted Kwakwaka'wakw artist, is by Yvette Cardozo, used here with her permission. The mask is from the private collection of Bill Hirsch and Yvette Cardozo. It is made of red cedar, 44 inches wide, 30 inches tall. When closed, the mask shows the head of a thunderbird. Strings are pulled to open the beak, showing both a human face in the center and sea monster designs.
19. Prevailing Kwakiutl resistance to the reduction of their living culture—the potlatch is enjoying a great resurgence among the canoe-faring First Nations peoples of the Northwest coast of Turtle Island—to museum artifacts provides, I believe, powerful testimony for Schelling's own plea for "a thoroughly indigenous art [*einer durchaus eigentümlichen Kunst*]" (I/7, 328). These transformational masks are inseparable from the soil that grants them their

life as cultural practices. On this issue, one can also see Deleuze and Guattari on the problem of earth in the fourth chapter ("Geophilosophy") of *What is Philosophy?* "The earth is not one element among others but rather brings together all the elements within a single embrace while using one or another of them to deterritorialize territory" (WP, 85).

20. Samuel Taylor Coleridge, *Biographia Literaria* (1817 edition), eds. James Engell and W. Jackson Bate (Princeton: Princeton University Press, 1983), 168.
21. Hermann Broch, "Evil in the Value-System of Art," *Geist and Zeitgeist: The Spiritual in an Unspiritual Age*, ed. and trans. John Hargraves (New York: Counterpoint, 2002), 37.
22. Milan Kundera, *The Unbearable Lightness of Being*, trans. Michael Heim (New York: Harper and Row, 1984), 256. For a discussion of the problem of kitsch with regard to Kundera in particular and as a problem more generally, see chapter 5 of my *Commiserating with Devastated Things: Milan Kundera and the Entitlements of Thinking* (New York: Fordham University Press, 2015).
23. Gilles Deleuze, *Cinema 2: The Time Image*, trans. Hugh Tomlinson and Robert Galeta (Minneapolis: University of Minnesota Press, 1989), 171–172.
24. All quotes from the 1797 System Fragment have been adapted, with revisions, from Diana I. Behler's translation, *Philosophy of German Idealism*, ed. Ernst Behler (New York: Continuum, 1987), 162–163. "Das sogenannte 'Älteste Systemprogram," *Materialien zu Schellings philosophischen Anfängen*, ed. Manfred Frank and Gerhard Kurz (Frankfurt am Main: Suhrkamp, 1975), 110–112.
25. Heidegger: "Whether nature is degraded to the exploitative place of calculation and furnishing [*Ausbeutungsgebiet des Rechnens und Einrichtens*] and to an opportunity to 'have an experience [*Erlebnis*]' or whether nature as the self-closing Earth bears the opening of a world without image." *Beiträge zur Philosophie (Vom Ereignis)*, ed. von Friedrich-Wilhelm von Herrmann (Frankfurt am Main: Vittorio Klostermann, 1989), 91. For a much needed and more balanced reappraisal of Schelling's relationship to Heidegger, see Lore Hühn's "A Philosophical Dialogue between Heidegger and Schelling," trans. David Carus, *Comparative and Continental Philosophy* 6.1 (Spring 2014), 16–34.

Chapter 6. Life of Imagination

1. G.W.F. Hegel, *Lectures on the History of Philosophy: The Lectures of 1825–1826: Volume Three: Medieval and Modern Philosophy*, trans. Robert Brown and J.M. Stewart with the assistance of H.S. Harris (Berkeley: University of California Press, 1990), 268.
2. Martin Heidegger, *Schellings Abhandlung über das Wesen der menschlichen Freiheit*, ed. H. Feick (Tübingen: Max Niemeyer, 1971), 4. *Schelling's Treatise On the Essence of Human Freedom*, trans. Joan Stambaugh (Athens: Ohio University Press, 1985), 3. I would also like to take this occasion to insist that Heidegger's demotion of Schelling to an inadvertent representative of metaphysics' ultimate culmination in the subjectivity of the will (and, as such, a precursor to Nietzsche's "will to power") is not only uncharitable, but it is also demonstrably false. In the Eschenmayer letter, for example, Schelling rails against the subjectivity that governs modern thought. "If followed consistently, all Kantian, Fichtean, and Jacobian philosophy, that is, the whole philosophy of subjectivity that rules our time, must come to your conclusion" (I/8, 167). Schelling's entire middle period was a concerted effort to battle against the very things of which Heidegger accused him. In this respect, see also Lore Hühn's "A Philosophical Dialogue between Heidegger and Schelling," trans. David Carus, *Comparative and Continental Philosophy* 6.1 (Spring 2014).
3. This is from the new forthcoming translation of *On the Deities of Samothrace* by David Farrell Krell in an edition currently under my editorship.
4. Among the Hawaiian peoples, for example, those who broke the taboos (*kapu*) or who were defeated at war, both punishable by death, were given a new lease on life if they made it to the Pu'uhonua or Place of Refuge. In the Pu'uhonua o Hōnaunau on the Big Island of Hawaii one can still feel the *mana* or natural potency, what Ernst Cassirer famously but not very delicately called "a common mysterious stuff that permeates all things." [Ernst Cassirer, *An Essay on Man: An Introduction to a Philosophy of Human Culture* (New Haven: Yale University Press, 1944), 96.] These are the very potencies into which the Cabiri were initiated.
5. See also IPU, 71.

6. *Denn Christentum ist Platonismus für's "Volk." Kritische Gesamtausgabe* 6:2, 4. This is from the preface to *Jenseits von Gut und Böse* (1885).
7. Adapted, with revisions, from Diana Behler's translation, *Philosophy of German Idealism*, ed. Ernst Behler (New York: Continuum, 1987), 162–163. For the German I consulted "Das sogenannte 'Älteste Systemprogram,'" *Materialien zu Schellings philosophischen Anfängen*, ed. Manfred Frank and Gerhard Kurz (Frankfurt am Main: Suhrkamp, 1975), 110–112. For an alternative translation, see also David Farrell Krell, *The Tragic Absolute: German Idealism and the Languishing of God* (Indiana: Indiana University Press, 2005), chapter 1.
8. See Krell, *The Tragic Absolute*, chapter 1. I regard this as the definitive analysis of the authorship controversy as well as the definitive dismissal of the importance of said controversy.
9. Quoted in Manfred Frank, *Der kommende Gott* (Frankfurt am Main: Suhrkamp, 1982), 34–35. This is precisely the iconic charge that Schelling made against Fichte (and his age) in his 1806 polemic, *Darlegung des wahren Verhältnisses der Naturphilosophie zu der verbesserten fichteschen Lehre*, where he explicitly accused Fichte of *Schwärmerei*. The *Schwärmer*, following Luther's condemnation of those who, claiming to have seen God, fanatically and uncritically swarmed into sects and schools, know what the ground is, and, in Fichte's case, posit nature outside of the subject, as something that resists the subject, but which should be brought under the subject's control. Schelling excoriated Fichte's thinking as *Bauernstolz* (I/7, 47), literally the self-congratulatory pride of a peasant who profits from nature without really grasping it. This lopsided and self-serving cultivation is at the heart of a contemporary nature annihilating *Schwärmerei*. "If an inflexible effort to force his subjectivity through his subjectivity as something universally valid and to exterminate all nature wherever possible and against it to make non-nature [*Unnatur*] a principle and to make all of the severity of a lopsided education in its dazzling isolation count as scientific truths can be called *Schwärmen*, then who in this whole era swarms in the authentic sense more terribly and loudly than Herr Fichte?" (I/7, 47).
10. Deleuze: "It is strange that aesthetics (as the science of the sensible) could be founded on what *can* be represented in the sensible.

True, the inverse procedure is not much better, consisting of the attempt to withdraw the pure sensible from representation and to determine it as that which remains once representation is removed (a contradictory flux, for example, or a rhapsody of sensations). Empiricism truly becomes transcendental, and aesthetics an apodictic discipline, only when we apprehend directly in the sensible that which can only be sensed, the being itself *of* the sensible: difference, potential difference and difference in intensity as the reason behind qualitative diversity. It is in difference that phenomena fulgurate and express themselves as signs (*fulgure, s'explique comme signe*), and movement is produced as an 'effect'" (DR, 56–57/79–80, translation altered).

11. On this issue I depart from Badiou and his insistence that for Deleuze, and by implication, Schelling, the "fundamental problem is most certainly not to liberate the multiple but to submit thinking to a renewed concept of the One" and a "metaphysics of the One" (10). Hence, despite Deleuze's "seemingly disparate cases" of analysis, his conceptual production in the end is "*monotonous*" (14), the ceaseless repetition of the same in which Deleuze must "refashion what he has already produced, and repeat his difference, in differentiating it even more acutely from other differences" (14). To the contrary: *the one can only be thought as the multiple.* Alain Badiou, *Deleuze: The Clamor of Being*, trans. Louise Burchill (Minneapolis: University of Minnesota Press, 2000).
12. Jean-Luc Nancy, "Why are There Several Arts?" *The Muses*, trans. Peggy Kamuf (Stanford: Stanford University Press, 1996), 2. Henceforth M.
13. *Hegel's Aesthetics: Lectures on Fine Arts*, volume one, trans. T. M. Knox (Oxford: Oxford University Press, 1975), 10.
14. John Sallis, *Transfigurements: On the True Sense of Art* (Chicago: University of Chicago Press, 2008), 154.
15. "and greatest" inexplicably dropped from Behler's translation.
16. "If love wanted to shatter the will of the ground, it would be contesting itself and not be at one with itself and would no longer be love. This letting be effective of the ground [*dieses Wirkenlassen des Grundes*] is the only thinkable concept of allowance [*Zulassung*], which in the customary relationship to human beings is utterly impermissible" (I/7, 375).
17. Das Dunkel der Schwere und der Glanz des Lichtwesens bringen

erst zusammen den schönen Schein des Lebens hervor, und vollenden das Ding zu dem eigentlich Realen, das wir so nennen.

18. Alain Badiou, *Being and Event*, trans. Oliver Feltham (London and New York: Continuum, 2005), 69.
19. See also John Sallis, *Force of Imagination* (Bloomington: Indiana University Press, 2000), 47–51.
20. The letter is found in Schelling, *Texte zur Philosophie der Kunst*, selected and edited by Werner Beierwaltes (Stuttgart; Reclam, 1982), 134–135.
21. Wallace Stevens, *The Necessary Angel: Essays on Reality and the Imagination* (originally published by Knopf in 1951), *Collected Poetry and Prose*, eds. Frank Kermode and Joan Richardson (New York: Library of America, 1997), 676, 737. Henceforth NA.
22. For more on this problem, see chapter 5.
23. See *Bruno, or, On the Divine and Natural Principle of Things* (1802), trans. Michael Vater (Albany: State University of New York Press, 1984), 231. See also Stott's helpful reference in his invaluable edition of *The Philosophy of Art* (1802–1804) (Minneapolis: University of Minnesota Press, 1989), 292 n.15. Although for reasons of stylistic consistency with the present work I have generally altered somewhat his translations, I have very much profited from them.
24. Jacques Lacan, "The Mirror Stage as Formative of the Function of the I as Revealed in Psychoanalytic Experience" (1949), *Écrits: A Selection*, trans. Alan Sheridan (New York and London: Norton, 1977), 6.
25. Sun Qianli, "Treatise on Calligraphy" (687), *Two Chinese Treatises on Calligraphy*, trans. Chang Ch'ung and Hans H. Frankel (New Haven and London: Yale University Press, 1995), 15.
26. For more on the question of language in relationship to Schelling, see my "The Language of Natural Silence: Schelling and the Poetic Word After Kant," *Kant and the Problem of Language: Essays in Criticism*, eds. Frank Schalow and Richard Velkley (Evanston: Northwestern University Press, 2014).
27. This translation is taken from *Historical-critical Introduction to the Philosophy of Mythology* (1842), trans. Mason Richey and Markus Zisselsberger (Albany: State University of New York Press, 2007), 167. Henceforth HCI.
28. Schelling located the phrase in a "wondrous" essay originally published in the *Transactions of the Royal Society for Literature*. "This

essay has particularly pleased me because it showed me how one of my earlier writings (the text on the *Deities of Samothrace*)—whose philosophical content and importance was little, or, rather, not at all understood in Germany—has been understood in its meaning by the talented Brit" (II/1, 196; HCI, 187).

29. See also Rudolf Hablützel, *Dialektik und Einbildungskraft: F. W. J. Schellings Lehre von der menschlichen Erkenntnis* (Basel: Verlag für Recht und Gesellschaft, 1954). The author characterizes Schelling's thinking as "a *mysticism of imagination* [*eine* Mystik der Einbildungskraft]" (81), although, as we shall see, this can be quite misleading and does not account for Schelling's own radical delimitation of mysticism as quietist and unable to know. I would speak of a mythology of imagination or an art of imagination or, best of all, a life of imagination.
30. As Schelling puts it in the Berlin *Historical-critical Introduction to the Philosophy of Mythology*: "Because mythology is not something that emerged artificially, but is rather something that emerged naturally—indeed, under the given presupposition, with necessity—*form* and *content*, *matter* and *outer appearance*, cannot be differentiated in it. The ideas are not first present in another form, but rather they emerge only in, and thus also at the same time with, this form" (II/1, 196; HCI, 136).
31. "The Mythological Being of Reflection—An Essay on Hegel, Schelling, and the Contingency of Necessity," Markus Gabriel and Slavoj Žižek, *Mythology, Madness and Laughter: Subjectivity in German Idealism* (London and New York: Continuum, 2009), 62. Henceforth MBR. On this critical issue, see also his important work, *Der Mensch im Mythos: Untersuchungen über Ontotheologie, Anthropologie und Sebstbewußtseinsgeschichte in Schellings* Philosophie der Mythologie (Berlin and New York: Walter de Gruyter, 2006).
32. Orrin F. Summerell formulates the problem thusly: "The symbol ('*Sinnbild*') unites the sensible and the intelligible by instantiating meaning ('*Sinn*') in a picture ('*Bild*'). . . . It manifests things as they really are, in the interiority of their original ideality, yet this itself in the form of a concrete object . . . a total mediation of sensibility and intelligibility through the real indifferentiation of meaning and being." "The Theory of the Imagination in Schelling's Philosophy of Identity," *Idealistic Studies* 34.1 (2004), 90.

33. See Daniel Whistler's important study, *Schelling's Theory of Symbolic Language: Forming the System of Identity* (Oxford: University of Oxford, 2013). "Language does not conceal its meaning behind a husk, nor does it retreat into the subject; in language, meaning takes on sensuous form without thereby reducing any of its meaningfulness. Language is meaningful being: it is tautegorical (and therefore synthetic)" (192). "Schelling is committed to an absolute interpretation of tautegory . . . meaning is utterly identical to being—there is nothing to distinguish them. . . . No distinction can be drawn between meaning and being; they exist equally and indissociably in a state of indifference . . . words cannot represent something in the world; in fact, there is no outside to words. Language, that is, is heautonomous" (198). Heautonomy, a term that is also found in Kant's third *Critique* to describe "a form of autonomous judgment that legislates only to itself" (15). Hence "the symbol is an image which need not look outside itself for meaning" (15). As I have argued in both this chapter and in the second chapter, this has critical implications for philosophical religion. Whistler: "This process could even be dubbed the de-absolutization of theology. Theology must be taken down from its pedestal and transformed into one more tool for poietic practice. . . . Only therefore by cutting theology down to size can all science, including theology itself, increase its productivity: theology's reformation is dependent on its simultaneous debasement" (242). We could even call this debasement the liberation of theology as it evolves into philosophical religion, its evolution from Petrine domination (the imposition of its form on all others), through Pauline negative theology (the intuition of the emptiness of form), to the A^3 of philosophical religion (see chapter 2), or what Schelling from the beginning called a new mythology of sensuous religion.
34. Georges Bataille, *The Accursed Share Volumes II and III*, trans. Robert Hurley (New York: Zone Books, 1991), 392.
35. Die Foderung einer Mythologie ist ja aber gerade, *nicht* daß ihre Symbole bloß Ideen bedeuten, sondern daß sie für sich selbst bedeudend, unabhängige Wesen seien.
36. Jeder originell behandelte Stoff is eben dadurch auch universell poetisch.
37. *The Letters of Herman Melville*, eds. Merrell R. Davis and William H. Gilman (New Haven: Yale University Press, 1960), 124–125.

38. Charles Olson, *Call Me Ishmael* (1947) (Baltimore and London: Johns Hopkins University Press, 1997), 102. Henceforth CMI.
39. Although I remain sympathetic to Marovitz's thesis, it is hard to typify Schelling or Melville as at all typical of a *Zeitgeist.* The utterly deaf ears on which *The Deities of Samothrace* fell or the fact that Melville was not sure that anyone beside Hawthorne would appreciate his book (indeed it was not widely read until after World War I) hardly a *Zeitgeist* make. That being said, the proximity of these two works testifies more clearly to the prescience of both.
40. Sanford Marovitz, "Melville's Problematic 'Being,'" *Bloom's Critical Modern Views: Melville*, new edition, ed. Harold Bloom (New York: Infobase, 2008), 50.
41. Kant makes this claim in a footnote in section 49 of the *Kritik der Urteilskraft* ("*Von den Vermögen des Gemüts*"). *Kritik der Urteilskraft* (1790), ed. Karl Vorländer (Hamburg: Felix Meiner Verlag, 1990), 171.

Appendix A

1. Schelling's letter originally appeared in I/8 of the Karl Schelling edition, and it appears in the fourth main volume of the Schröter edition (*Schriften zur Philosophie der Freiheit*). To facilitate a study of this text with the German original, we have interpolated the page numbers of the standard citation, preserved in both of these editions. All of the notes, with the exception of one (the beginning of note number 13, which was inserted by Schelling at a later date), are the joint responsibility of the translators.
2. *Das Wesen* is the German word for essence, but one should proceed with caution in the sense that Schelling does not associate this term with its *Wirkungsgeschichte* in Platonism where it names transhistorical forms (ideas) that have their a priori and intelligible being in and for themselves and instantiate themselves derivatively in the sensuous realm of becoming. (Schelling's own reading of the *Phaedrus* in this letter speaks against this reading.) For Schelling, *das Wesen* names the tension within the complete presentation of the times, between present being (existence) and the simultaneous intimation of that which is as no longer being (the past) and that which is as not yet being (the future).

3. This is a reference to the famous phrase, *der nie aufgehende Rest* in the *Freedom* essay. "Following the eternal deed of Self-Revelation, the world is how we now perceive it, namely, everything is rule, order, and form. But the unruly [*das Regelose*] always still lies within Ground as if it could once more again break through. Nowhere does it appear as if order and form were originary but rather as if an initial Unruliness had been brought to order. This is the incomprehensible basis of reality within things, the irreducible remainder [*der nie aufgehende Rest*], which remains eternally within Ground and which with the greatest effort does not admit of being resolved in the understanding. The understanding [*Verstand*] in its actual sense is born out of this incomprehensibility [*Verstandlose*]. Without this preceding darkness there is no reality of creatures. Darkness is their necessary inheritance" (I/7, 360). *Aufgehen* will be translated as "resolve" in the sense of absorbing or assimilating one matter into another without leaving a remainder or a surplus or excess, something that retained itself while at the same time becoming its opposite. A=B does not imply that A and B, considered in themselves, do not also contest each other. They do not become one and the same (*einerlei*). *Auflösen*, on the other hand, means to dissolve and will be used in this sense.
4. For a discussion of the problem of the μὴ ὄν, see chapter 3 of this book. There is a notorious confusion in handling the distinction between *das Seyn* and *das Seyende*. The latter is a noun made from the verb to be, is a particular being, that which, as Schelling makes clear later in the letter, exists. The *Nicht-Seyende* is that within being that does not have being. *Die Weltalter*: "This extremity can itself be called only a shadow of the being, a minimum of reality, only to some extent still having being, but not really. This is the meaning of non-being according to the Neo-Platonists, who no longer understood Plato's real meaning of it. We, following the opposite direction, also recognize an extremity, below which there is nothing, but it is for us not something ultimate, but something primary, out of which all things begin, an eternal beginning [*ein ewiger Anfang*], not a mere feebleness or lack in the being, but active negation" (1/8, 245).
5. The passage from Eschenmayer's October 1810 letter to which Schelling is responding runs as follows: "First, I am hung up on the principle of your entire deduction. You say that since nothing

is prior to or outside of God, he must therefore have the ground of his existence in himself, and then you distinguish this ground from God insofar as He exists. Against this, I contend that if God has the ground of his existence in Himself, the ground immediately ceases to be a ground and collapses into one with His existence. In the sequence of natural things, we can indeed distinguish between ground and consequence, substance and accident, etc. We can distinguish a body at rest from its movement, an organ from its function, or even a decision of the will from its externalization, but for God this distinction is invalid, because even under your assumption ground and consequence, form and essence, and being and becoming collapse into a single point. [146] My contention is not only that ground cannot be a predicate of God, but the entire logical mode of thinking gives no measure for God, and in the application of our concepts of ground and consequence, form and essence, being and becoming, He is denigrated to an object of the understanding [*Verstandeswesen*] that can take up no other role in the imagined sphere that we attribute to him than the ego [*unsere Ich*] can in its sphere. Compared with your earlier claims [cf. Schelling's 1804 *System of Philosophy in General* (I/6, 148)—Trans.], this assumption seems like a step backwards. In the journals this means that reason does not have the idea of God but is this idea itself and nothing else. In this proposition in which reason is posited as the idea of God, the understanding's concepts of ground and consequent, etc., can find no more application because unity does not allow totality its proper measure, and reason as the universal organ cannot be absorbed back into the understanding as the particular organ. The God that you so posit is to me merely a particular God, like our ego, and all constructions that you would have begin with Him, which however are so positioned as if they were His own, have no higher consequence than the human ego has in its own system" (I/8, 145–146).

6. God as a being could not logically follow from or be entailed by the *Urwesen* of God as the μὴ ὄν, that which in being does not have being.
7. *Geht über die Erde hinaus*: go through in order to get beyond. The earth is already the thought of God.
8. Eschenmayer writes: "You give God the longing to give birth to Himself, as if there could be in God a wish to be something that

He is not already. When the will incorporates the mind (*Gemüth*), there arises what we call longing, an entirely human process that does not allow of any extension to God. Likewise, the same could be said of *personality, autonomy, self-consciousness, self-cognition, life, etc.* These are all a mixture of the free principle with the necessary one and thus purely human and inadequate to God's dignity" (I/8, 148).

9. To *conflate* is to *vereinerleien*, to make one and the same thing, a verb that stems from *einerlei*, to be one and the same thing. It reduces the duality of ground and existence to a unity that eliminates their internal contestation. It conflates or confuses the ground of God/Being as such with a being or beings.
10. We have revised Schelling's citations of his own works to indicate the pagination in the standard edition of his collected works.
11. Here Schelling seems to have been careless in his transcription of Eschenmayer, since the original quotation uses "is" (*ist*), not "was" (*war*). And since, whatever its form, his use of the verb "to be" is a hypothetical one aimed at criticizing Schelling's use of the verb, it is not clear that Schelling's cry of hypocrisy is justified. Note, though, that Schelling uses the indicative *was* and not the contrary-to-fact subjunctive *were*.
12. Eschenmayer's objection reads: "According to your view, the understanding arises from lack of understanding, order from chaos, [153] and light from the dark ground of gravity. Is there anything to hinder us from taking these oppositions further and allowing virtue to arise from vice, saintliness from sin, heaven from hell, and God from the devil? For what you call the dark ground of existence is indeed something similar to the devil. Now in these oppositions there is a gradual development toward the light, a rising of the innermost center, etc. Your view here agrees with the ancient myths which have their first god born out of the blind night. But I ask, to what end [is] all of this? If God is from eternity, how should the lack of understanding precede understanding, chaos precede order, or darkness precede light? The eternal indeed cannot participate in time, nor can it abide any change in things in themselves. Just as little can I accept even analogously applying the relationship of natural principles like light and gravity to the ethical order of things. Good and evil first arise, then, when the natural principles have entirely vanished and lost all of their meaning. The human

being is thus free because no natural law has compulsion [*Zwang*] over him" (I/8, 152–3).

13. For example, Süskind. See the *Proof of the Schellingian Doctrine of God* in the seventeenth issue of the *Magazine for Christian Morals and Dogmatics* (in a special imprint with Cotta in Tübingen). What it has to do with this (alleged) proof, I will take the opportunity to indicate thoroughly in a future volume. [Comment inserted at a later time]—Schelling. [Schelling's later inserted note is referring to a vitriolically polemical work published by the one-time Tübingen theology professor, Friedrich Gottlieb Süskind (February 17, 1767–November 12, 1829), called *Prüfung der Schellingischen Lehren von Gott, Weltschöpfung, Freyheit, moralischem Guten und Bösen* (Tübingen: In der J. G. Cotta'schen Buchhandlung, 1812). (This refers to the reprint of the work in book form with the same press.) He had been a member of what the young Schelling, Hegel, and Hölderlin had derided as *die Tübinger Orthodoxie*. During this period of Schelling's life, the attacks against his thinking continued to mount, including not only Adam Karl August von Eschenmayer and Friedrich Gottlieb Süskind, both of whom in their own ways are defending their commitments to supernaturalism (what Eschenmayer associates with "nonphilosophy"), but also his Munich colleague (and former friend), Friedrich Heinrich Jacobi, whose *Von den göttlichen Dingen und ihrer Offenbarung*, appeared in 1811, and to which Schelling published a spirited response.]
14. Despite the quotation marks, this is not a direct quote. Schelling is responding primarily to this passage of Eschenmayer's letter: "To make myself as clear as possible, your treatment of human freedom seems to me a complete transformation of ethics into physics, a swallowing up of the free by the necessary, of the soul by the understanding, of the ethical by the natural, and most of all a complete depotentiation of the higher order of things into the lower. Where are duty, law, conscience, and virtue to be found in your system? They do not emerge from proportion and disproportion, as between health and sickness, for these are characteristic of the level of the organic, not the ethical, and even less do they emerge from that dynamical progression of degrees [*Gradreihe*], where two forces separate from a common center and at last stand over against each other as light and darkness. The center and its periphery, which play the principal roles on the side of nature, no longer find an application on the ethical side. The more the true butts

up [*fortrückt*] against the beautiful, or the purely physical against the organic, the more must the circle be broken (even viewed as a symbol) and one or two sides open up to the infinite as in a parabola or hyperbola. But in the ethical order—where this is no longer valid, and only the transcendent lines of infinite order, which give an uncountable number of y's for every x—a noble image of willing and freedom—an analogy is still revealed to us in the ethical world. The history of humanity is a cycloid" (I/8, 150).

15. We have replaced the pagination of Schelling's original edition with the Stephanus pagination.
16. The "self-deception" Schelling is criticizing appears in this passage: "For me reason is the universal organ in which the totality exclusively resides and the understanding the particular organ in which unity [*Einheit*] holds sway, and sense [*der Sinn*] the singular organ to which alone multiplicity belongs. There is not enough space here to circumscribe every organ and to show how each grasps the others, and what each characteristically is. I note simply that the individuality of the ego appears as the absolute unity of these three systems, but as such has its center not in the universal organ, as philosophy hitherto assumed, but in the particular organ, namely in the understanding. The ego can never leave its unity; it can never elevate itself to totality, nor sink to multiplicity. This unity has always been confused with absolute identity, [147] which is entirely impermissible; even so it is an error of our [German] language that we have not learned to distinguish between individuality [*Einzelnheit*], which only indicates one of many, and a unity [*Einheit*] in which multiplicity is itself tied up" (I/8, 146–147).
17. *Das Ebenbild* is literally the precise or exact image, or in more colloquial English, the "spitting image." It is used here in reference to Genesis 1:26–27: "Then God said, 'Let us make mankind in our image, in our likeness, so that they may rule over the fish in the sea and the birds in the sky, over the livestock and all the wild animals, and over all the creatures that move along the ground.' So God created mankind in his own image, in the image of God he created them; male and female he created them" (New International Version).
18. Schelling is here playing on two senses of *Einfall.* In the first sense it means a "sudden thought," to come or "fall" to thought, that is, for a thought to occur to thinking (it occurred to us to make ourselves in the image of God). For Schelling, we were made in the

image of God, but the loss of this relationship speaks to an ancillary meaning of *Einfall*, namely, the Fall of original sin (in the myth of Eden). One can speak of the *der Einfall der Nacht*, the fall of night, but here the occurrence is the sudden Fall from grace, or what is more typically called *der Sündenfall.* Or as the *Freedom* essay has it, in the abandonment of the center for the periphery, it has occurred to us abandon ourselves as in the image of God, to flee ourselves as *Gegenbilder* of the divine *Urbild.*

19. Schelling is quoting from parts of the opening two paragraphs of Nicholas Malebranche's *The Search after Truth* (first edition, 1674, final edition 1712). We have taken the translation from the translation by Thomas M. Lennon and Paul J. Olscamp (Columbus: Ohio State University Press, 1980), xix. Because there are a few striking discrepancies between the Lennon/Olscamp translation from the French and Schelling's German version, we have marked them in the text by including Schelling's German at the appropriate points.
20. The *Mitwissern* or co-knowers alludes to Schelling's *Mitwissenschaft*, co-knowing, which literally translates the Latin *conscientiæ*: knowing or knowledge (*scientia*) in an ancillary and joint fashion (*con*). One can detect at least three of the senses that Schelling unfolds in Latin *conscientiæ/Mitwissenschaft*: joint knowledge, consciousness, as well as the ethical sense of the conscience. In the third draft of *Die Weltalter*, we read: "Created out of the source of things and the same as it, the human soul is the *Mitwissenschaft* of creation. In the soul lies the highest clarity of all things, and the soul is not so much knowing as knowledge itself" (I/8, 200).
21. This is an allusion to Ephesians 2:14: "For he himself is our peace, who has made the two one and has destroyed the barrier, the dividing wall of hostility" (*New International Version*).
22. Schelling's intermediary concepts (*Mittelbegriffe*), which he often claims to be his most important and what contemporary philosophy most severely lacks (I/2, 150), are the concepts (A^3) that name the productive holding together of the opposites.

Appendix B

1. To be fair, Eschenmayer's own letter virtually invites Schelling's condescension with a healthy dose of arrogance and self-righteousness.

Although there was not room to include a full translation of Eschenmayer's letter in this book, a translation of the complete letter can be found on the website of the North American Schelling Society (NASS).

2. All references to Schelling follow the standard citation format used throughout this book.
3. Xavier Tilliette, *Schelling: Biographie*, translated into German by Susanne Schaper (Stuttgart: Klett-Cotta, 2004), 260.
4. The full passage reads: "To express myself completely regarding what I mean by the system of reason and what I mean by God, I must frankly deny much of what you say about him in your essay. God has no nature, no ground in himself. 'In-himself' and 'outside-himself' have no meaning for God; there is no ground that acts independently of God, which for you contains the possibility of the evil principle. God is not a unity and is definitely not divisible into two principles that separate in human beings, which, to put it casually, denies the freedom which is based upon this division. There are not two equally eternal beginnings from a single indifference, for the eternal is not one and two, but the all" (I/8, 150).
5. This passage could provide evidence that even before their famous correspondence, Schelling viewed Eschenmayer's conception of nature as inadequate. Jantzen names Eschenmayer (along with Jacobi) as a possible target of this polemic (Jörg Jantzen, "Die Möglichkeit des Guten und des Bösen," [350–364], in *F.W.J. Schelling: Über das Wesen der menschlichen Freiheit*, eds. Otfried Höffe and Annemarie Pieper, [Berlin: Akademie Verlag, 1995], 77n).
6. "*Urwesen*" appears only twice in the *Freedom* essay, at I/7, 337 and I/7, 354.
7. Various authors have explained the problems with translating Schelling's *Wesen* as "essence," including Love and Schmidt ("Translators' Note" in Schelling, *Philosophical Investigations into the Essence of Human Freedom*, Albany: State University of New York Press, 2006), Krell (*The Tragic Absolute*, Bloomington: Indiana University Press, 81–84) and Wirth (*The Conspiracy of Life*, Albany: State University of New York Press, 2003, 158–160), and Wirth even goes so far as to render it as "being" in all cases (Schelling, *The Ages of the World*, trans. Wirth, Albany: State University of New York Press, 2000, xxxi). Although this is undoubtedly preferable for such terms as *Lebewesen*, "living being," God's *Urwesen* is defined

precisely by its indifference to the contradictions necessary for life, making the more abstract "essence" preferable. (See also note 2 in the preceding translation of Schelling's Eschenmayer Letter.)

8. By my count, *Verstand* appears as a noun 48 times (excluding compounds) and *Vernunft* 31 times.
9. Schelling, *Ideas for a Philosophy of Nature*, trans. E. E. Harris and Peter Heath (Cambridge University Press, 1988), 10 (I/2, 13) and 160 (I/2, 200). Cf. the 1796–1797 *Treatises Explicatory of the Idealism in the Wissenschaftslehre*, where Schelling calls the understanding a "secondary, derivative, ideal faculty" (I/1, 422).
10. Eschenmayer writes, "There are not two equally eternal beginnings from a single indifference, for the eternal is not one and two, but the all. There are admittedly three ideas, namely truth, beauty, and virtue, with which everyone represents the totality of our universal organ in some way, and in which the eternal itself appears divided, only the indeterminacy of our language calls here for a supplement. The eternal rests only in the harmony of those ideas, but taken for themselves they express simply the orders of the infinite, so that truth indicates the lower order or the purely physical, beauty the middle order or the organic, and virtue the higher order or the ethical" (I/8, 150).
11. In the *Presentation of My System of Philosophy*, for instance, Schelling argues that reason is identical not only with philosophy, but with everything conceivable, and hence uses the phrases "in itself [*an sich*]" and "in reason" interchangeably (I/4, 115).
12. See, for example, *On the World-Soul* (1798), where Schelling states that a person who learns to see every individual thing as an expression of God, who realizes that "the violent drive toward determination is undeniably in all metals and stones, in the immeasurable power of which all existence is an expression," surrenders all hopes to grasp nature through the understanding and "at last enters reason, the holy Sabbath of nature, where it, at peace with its earlier works, recognizes and construes [*erkennt und deutet*] itself as itself" (I/2, 378).
13. In the *Stuttgart Seminars*, Schelling makes the same point by calling both evil and disease "nonbeings [*Nichtseyendes*]": "A nonbeing frequently impresses on us as a being, when seen from another perspective. What, for example, is disease? A state that is *counter to* nature, consequently a state that could not *be* and nevertheless is, a

state that has no real ground and yet possesses undeniably a fearsome [*furchtbare*] reality. Evil is for the moral world what disease is for the physical world; viewed in one way, it is the most definitive nonbeing while possessing a terrible [*schreckliche*] reality in another way" (I/7, 436–7).

14. Hegel, *Phänomenologie des Geistes, Gesammelte Werke*, vol. 9, ed. Nordrhein-Westfälischen Akademie der Wissenschaften (Hamburg: Felix Meiner Verlag, 1971–2006), 27, ¶32.
15. For a concise discussion of this distinction as it plays out in the *Freedom* essay and *Ages of the World*, see Slavoj Žižek, *The Abyss of Freedom* (Ann Arbor: University of Michigan Press, 1997), 7–14.
16. Cf. *Erlanger Vorträge* (I/9, 241).
17. *Kritik der Urtheilskraft, Kants Gesammelte Schriften*, vol. 5, ed. *Königlich Preußische Akademie der Wissenschaften* (Berlin: Walter de Gruyter, 1902–1942), 175.
18. Martin Heidegger, *Schelling's Treatise on the Essence of Human Freedom*, trans. Joan Stambaugh (Athens: Ohio University Press, 1985), 163. Henceforth ST.
19. See, for instance, the 1799 *First Outline of a System of the Philosophy of Nature*, trans. Keith Peterson (Albany: State University of New York Press, 2004), 5.

BIBLIOGRAPHY

Schelling

Schellings Werke: Nach der Originalausgabe in neuer Anordnung. Edited by Manfred Schröter. Munich: C. H. Beck. Printed twice: 1927–1959 and 1962–1971.

Citations in the present work follow the standard pagination, which follows the original edition of two divisions and fourteen volumes established by his son, Karl, after Schelling's death. It lists the division, followed by the volume, followed by the page number. Hence, (I/1, 1) would read, division one, volume one, page one. It is preserved in Manfred Schröter's critical reorganization of this material, although Schröter numbers the volumes without division from I-XIV, so a citation that might read SW XI would be for our present purposes II/1.

Werke: Historisch-Kritische Ausgabe. Edited by Hans Michael Baumgartner, Wilhelm G. Jacobs, Hermann Krings, Hermann Zeltner. Stuttgart (Bad Cannstatt): Frommann-Holzboog, 1976–present.

This is an ongoing and impressively rigorous project and it is virtually assured that this edition will not be completed in my lifetime.

Individual Editions Consulted

The Ages of the World (1813 draft). Translated by Judith Norman. *The Abyss of Freedom/Ages of the World*. Ann Arbor: University of Michigan Press, 1997. Includes an interpretive essay by Slavoj Žižek.

The Ages of the World (1815 draft). Translated by Frederick de Wolfe Bolman Jr. New York: Columbia University Press, 1942. (Reprinted: New York: AMS Press, 1967.)

The Ages of the World (1815 draft). Translated by Jason M. Wirth. Albany: State University of New York Press, 2000.

Aus Schellings Leben: In Briefen, 3 volumes. Edited by G.L. Plitt. Leipzig: S. Hirzel, 1869–1870.

Brief über den Tod Carolines vom 2. Oktober, 1809. Edited by Johann Ludwig Döderlein. *Kleine kommentierte Texte I*. Stuttgart (Bad Cannstatt): Frommann-Holzboog, 1975.

Briefe und Dokumente, vol. 1: 1775–1809. Edited by Horst Fuhrmans. Bonn: H. Bouvier Verlag, 1962.

Bruno, or, On the Divine and Natural Principle of Things (1802). Translated by Michael Vater. Albany: State University of New York Press, 1984.

Clara: Über den Zusammenhang der Natur mit der Geisterwelt (1810). Edited by Konrad Dietzfelbinger. Andechs: Dingfelder Verlag, 1987. *Clara.* Translated by Fiona Steinkamp. Albany: State University of New York Press, 2002.

The Deities of Samothrace (1815). Translated by Robert Brown. Missoula, MT: Scholars Press, 1977.

Einleitung in die Philosophie (1830). Edited by Walter Ehrhardt. Stuttgart (Bad Cannstatt): Frommann-Holzboog, 1989.

The Endgame of Idealism. Translated by Thomas Pfau. Albany: State University of New York Press, 1996. Collects together three essays: *Treatise Explicatory of the Idealism in the* **Science of Knowledge** (1797), *System of Philosophy in General and of the Philosophy of Nature in Particular* (1804), and the *Stuttgart Lectures* (1810).

Essais. Translated by Samuel Jankélévitch. Paris: Aubier, 1946.

Fichte-Schelling Briefwechsel. Edited by Walter Schulz. Frankfurt am Main: Suhrkamp, 1968.

First Outline of a System of the Philosophy of Nature (1799). Translated by Keith R. Peterson. Albany: State University of New York Press, 2004.

F.W.J. Schelling: Briefe und Dokumente, vol. I: 1775–1809. Edited by Horst Fuhrmans. Bonn: H. Bouvier, 1962.

F.W.J. Schelling: Briefe und Dokumente, vol. II: 1775–1803, *Zusatzband*. Edited by Horst Fuhrmans. Bonn: Bouvier Verlag Herbert Grundmann, 1973.

The Grounding of Positive Philosophy: The Berlin Lectures (1842–1843). Translated by Bruce Matthews. Albany: State University of New York Press, 2007.

Grundlegung der positiven Philosophie: Münchener Vorlesung WS 1832/33 und SS 1833. Edited by Horst Fuhrmans. Turin: Bottega d'Erasmo, 1972.

Historical-critical Introduction to the Philosophy of Mythology (1842). Translated by Mason Richey and Markus Zisselsberger. Albany: State University of New York Press, 2007.

Ideas for a Philosophy of Nature (1797). Translated by Errol E. Harris and Peter Heath. Cambridge: Cambridge University Press, 1988.

Initia Philosophiæ Universæ (1820–1821). Edited with Commentary by Horst Fuhrmans. Bonn: H. Bouvier, 1969. This is an early critical edition of the *Erlanger Vorlesungen*. These lectures formed the basis of Schelling's essay, *On the Nature of Philosophy as Science* (1821). A new and more complete version is now in the works.

Materialien zu Schellings philosophischen Anfängen. Edited by Manfred Frank and Gerhard Kurz. Frankfurt am Main: Suhrkamp, 1975.

On Dante in Relation to Philosophy (1803). Translated by Elizabeth Rubenstein and David Simpson. *The Origins of Modern Critical Thought: German Aesthetic and Literary Criticism from Lessing to Hegel*. Edited by David Simpson. Cambridge: Cambridge University Press, 1988, 239–247.

On the Deities of Samothrace. Edited by Jason M. Wirth and translated by David F. Krell, forthcoming.

On the History of Modern Philosophy (1827). Translated by Andrew Bowie. Cambridge: Cambridge University Press, 1996.

On the Nature of Philosophy as Science (1821). Translated by Marcus Weigelt. In *German Idealist Philosophy*. Edited with an Introduction by Rüdiger Bubner. London and New York: Penguin Books, 1997, 210–243.

"On the Relation of the Ideal and the Real in Nature, or Development of the First Principles of Naturephilosophy from the Principles of Gravity and Light [*Über das Verhältniß des Realen und Idealen in der*

Natur, oder Entwickelung der ersten Grundsätze der Naturphilosophie an den Principien der Schwere und des Lichts]." Addendum [*Zugabe*] to *Von der Weltseele*. Unpublished translation by Iain Hamilton Grant, 2013.

On the Source of the Eternal Truths (1850). Translated by Edward Allen Beach. *Owl of Minerva* (Fall 1990), 55–67.

The Philosophical Rupture Between Fichte and Schelling. Edited and translated by Michael Vater and David W. Wood. Albany: State University of New York Press, 2012.

Philosophische Entwürfe und Tagebücher 1809–1813: Philosophie der Freiheit und der Weltalter. Edited by Lothar Knatz, Hans Jörg Sandkühler, and Martin Schraven. Hamburg: Felix Meiner Verlag, 1994.

Philosophische Untersuchungen über das Wesen der menschlichen Freiheit und die damit zusammenhängenden Gegenstände (1809). The text originally appeared as the fifth and final essay in F.W.J. Schellings *philosophische Schriften*, vol. 1 (no other volumes appeared). Landshut: Philipp Krüll, *Universitätsbuchhandler*, 1809, 397–511. It was later included in the Werke, edited by Schelling's son, Karl Schelling. There is now an invaluable critical edition edited by Thomas Buchheim. Hamburg: Felix Meiner Verlag, 1997. There are three English translations. The first is by James Gutmann. La Salle, IL: Open Court, 1936. The second is by Priscilla Hayden-Roy. *Philosophy of German Idealism*. Edited by Ernst Behler. New York: Continuum, 1987, 217–284. The third and most recent is *Philosophical Investigations into the Essence of Human Freedom* (1809). Translated by Jeff Love and Johannes Schmidt. Albany: State University of New York Press, 2006. This includes a valuable translation of some of the background sources for the *Freedom* essay.

Philosophie der Mythologie: Nachschrift der letzten Münchner Vorlesungen 1841. Stuttgart (Bad Cannstatt): Frommann-Holzboog, 1996.

Philosophie der Offenbarung (1841–1842). Edited with an Introduction by Manfred Frank. Frankfurt am Main, 1977. This is an edition (supplemented with *Materialien*) of: *Die endlich offenbar gewordene positive Philosophie der Offenbarung*. Edited (with substantial polemical commentary) by H.E.G. Paulus. Darmstadt: Carl Wilhelm Leske, 1843.

Philosophy and Religion (1804). Translated by Klaus Ottmann. New York: Spring, 2010.

The Philosophy of Art (1802–04). Edited and translated by Douglas Stott. *The Theory and History of Literature*, vol. 58. Minneapolis: University of Minnesota Press, 1989.

Schelling im Spiegel seiner Zeitgenossen. Edited by Xavier Tilliette. Turin: Bottega d'Erasmo, 1974.

Schelling im Spiegel seiner Zeitgenossen, Ergänzungsband. Edited by Xavier Tilliette. Turin: Bottega d'Erasmo, 1981.

Schellingiana Rariora. Edited by Luigi Pareyson. Turin: Bottega d'Erasmo, 1977.

"Schelling's Aphorisms of 1805," *Idealistic Studies* 14 (1984), 237–258.

Schelling's Philosophy of Mythology and Revelation: Three of Seven Books. Translated and reduced by Victor C. Hayes. Erskineville, Australia: Australian Association for the Study of Religions, 1995.

Stuttgarter Privatvorlesungen (1810), unedited version. Annotated by Miklos Vetö. Turin: Bottega d'Erasmo, 1973. English translation published in *The Endgame of Idealism*. Translated by Thomas Pfau. Albany: State University of New York Press, 1996.

System des transzendentalen Idealismus (1800). Edited by Horst D.Brandt and Peter Müller. Hamburg: Felix Meiner, 1992. *System of Transcendental Idealism*. Translated by Peter Heath. Charlottesville: University of Virginia Press, 1978.

System der Weltalter: Müchener Vorlesung 1827–28. Edited by Siegbert Peetz. Frankfurt am Main: Vittorio Klostermann, 1990.

Das Tagebuch 1848. Edited by Hans-Jörg Sandkühler. Hamburg: Felix Meiner, 1990.

Texte zur Philosophie der Kunst. Selected and edited by Werner Beierwaltes. Stuttgart: Reclam, 1982.

Timaeus: Ein Manuskript zu Platon (1794). Edited by Harmut Buchner. Stuttgart (Bad Cannstatt): Frommann-Holzboog, 1995.

Über das Verhältnis der bildenden Künste zu der Natur (1807). Edited by Lucia Sziborsky. Hamburg: Felix Meiner Verlag, 1983. *The Philosophy of Art: An Oration on the Relation between the Plastic Arts and The Nature*. Translated by A. Johnson. London: John Chapman, 1845. A second translation by Michael Bullock was published as an appendix to Herbert Read. *The True Voice of Feeling: Studies in English Romantic Poetry*. New York: Pantheon Books, 1953.

The Unconditional in Human Knowledge: Four Early Essays (1794–1796). Translated by Fritz Marti. Lewisburg: Bucknell University Press, 1980.

Urfassung Philosophie der Offenbarung, 2 volumes. Edited by Walter E. Ehrhardt. Hamburg: Felix Meiner Verlag, 1992.

Vorlesungen über die Methode (Lehrart) des akademischen Studiums (1803), 2nd edition. Edited by Walter E. Ehrhardt. Hamburg: Felix Meiner Verlag, 1990. *On University Studies*. Translated by E.S. Morgan. Athens: University of Ohio Press, 1966.

Die Weltalter in den Urfassungen von 1811 und 1813 (Nachlaßband). Edited by Manfred Schröter. Munich: Beck, 1946.

Other Materials Consulted

Addiss, Stephen, with Stanley Lombardo and Judith Roitman. *Zen Sourcebook: Traditional Documents from China, Korea, and Japan*. Indianapolis and Cambridge: Hackett, 2008.

Allwohn, Adolf. *Der Mythos bei Schelling*. *Kant-Studien*, Ergänzungsheft. Charlottenburg: Pan-Verlag Rolf Heise, 1927.

Angehrn, Emil. *Die Frage nach dem Ursprung: Philosophie zwischen Ursprungsdenken und Ursprungskritik*. Munich: Wilhelm Fink, 2007.

Baader, Franz Xaver von. *Gesammelte Schriften zur Naturphilosophie*. Edited by Franz Hoffmann. Aalen: Scientia Verlag, 1963.

———. *Gesammelte Schriften zur philosophischen Erkenntniswissenschaft als spekulative Logik*. Aalen: Scientia Verlag, 1963.

Badiou, Alain. *Being and Event*. Translated by Oliver Feltham. London and New York: Continuum, 2005.

———. *Deleuze: The Clamor of Being*. Translated by Louise Burchill. Minneapolis: University of Minnesota Press, 2000.

Bataille, Georges. *The Accursed Share*, vol. 1. Translated by Robert Hurley. New York: Zone Books, 1991.

———. *The Accursed Share*, vol. 2 and 3. Translated by Robert Hurley. New York: Zone Books, 1993.

———. *Theory of Religion*. Translated by Robert Hurley. New York: Zone Books, 1989.

Baum, Klaus. *Die Transzendierung des Mythos zur Philosophie und Ästhetik Schelling und Adornos*. Würzburg: Königshausen und Neumann, 1988.

Baumgartner, Hans Michael, and Jacobs, Wilhelm G., eds. *Philosophie der Subjektivität? Zur Bestimmung des neuzeitlichen Philosophierens:*

Akten des 1. Kongresses der Internationalen Schelling-Gesellschaft 1989. Schellingiana, vol. 2. Stuttgart (Bad Cannstatt): Frommann-Holzboog, 1993.

———. *Schellings Weg zur Freiheitsschrift: Legende und Wirklichkeit: Akten der Fachtagung der Internationalen Schelling-Gesellschaft, 1992. Schellingiana*, vol. 5. Stuttgart (Bad Cannstatt): Frommann-Holzboog, 1996.

Beach, Edward Allen. *The Potencies of God(s): Schelling's Philosophy of Mythology*. Albany: State University of New York Press, 1994.

Beiser, Frederick C. *The Fate of Reason: German Philosophy from Kant to Fichte*. Cambridge: Harvard University Press, 1987.

———. *German Idealism: The Struggle Against Subjectivism 1781–1801*. Cambridge: Harvard University Press, 2002.

Benz, Ernst. *Schellings theologische Geistesahnen*. Wiesbaden: Akademie der Wissenschaften und der Literatur, in Kommission bei Franz Steiner Verlag, 1955.

———. *Schelling: Werden und Wirken seines Denkens*. Zürich and Stuttgart: Rhein-Verlag, 1955.

Bergson, Henri. *Matter and Memory* (1896). Translated by N.M. Paul and W.S. Palmer. New York: Zone Books, 1988.

Betz, John R. "Reading 'Sibylline Leaves': J.G. Hamann in the History of Ideas." *Hamann and the Tradition*. Edited by Lisa Marie Anderson. Evanston: Northwestern University Press, 2012, 5–32.

Beierwaltes, Werner. "The Legacy of Neoplatonism in F.W.J. Schelling's Thought." Translated by Peter Adamson. *International Journal of Philosophical Studies* 10.4 (2002), 393–428.

Bloch, Ernst. *Erbschaft dieser Zeit*. Frankfurt am Main: Suhrkamp, 1962.

———. *Das Materialismus Problem: Seine Geschichte und Substanz*. Frankfurt am Main: Suhrkamp, 1972.

Borges, Jorge Luis. "A Defense of *Bouvard et Pécuchet*" (1954), *Selected Non-Fictions*. Edited by Eliot Weinberger. New York: Viking, 1999.

Bowie, Andrew. *Aesthetics and Subjectivity: From Kant to Nietzsche*. Manchester: Manchester University Press, 1990.

———. *Schelling and Modern European Philosophy: An Introduction*. London and New York: Routledge, 1993.

Breton, Stanislas. *The Word and the Cross*. Translated by Jacquelyn Porter. New York: Fordham University Press, 2002.

Brito, Emilio. *Philosophie et théologie dans l'oeuvre de Schelling*. Paris: Cerf, 2000.

Broch, Hermann. *The Death of Virgil* (1945). Translated by Jean Starr Untermeyer. New York: Vintage International, 1995.

———. "Evil in the Value-System of Art." *Geist and Zeitgeist: The Spiritual in an Unspiritual Age*. Edited and translated by John Hargraves. New York: Counterpoint, 2002.

Brown, Robert F. *The Later Philosophy of Schelling: The Influence of Böhme on the Works of 1809–1815*. Lewisburg: Bucknell University Press, 1977.

Buchheim, Thomas. *Eins von Allem: Die Selbstbescheidigung des Idealismus in Schellings Spätphilosophie*. Hamburg: Felix Meiner Verlag, 1992.

———. "Zur Unterscheidung von negativer und positiver Philosophie beim späten Schelling." *Berliner Schelling Studien 2: Vorträge der Schelling-Forschungsstelle Berlin*. Edited by Elke Hahn. Berlin: Total Verlag, 2001, 125–145.

———, and Friedrich Hermanni, eds. "Alle Persönlichkeit ruht auf einem dunkeln Grunde": *Schellings Philosophie der Personalität*. Berlin: Akademie Verlag, 2004.

Büchner, Georg. *Lenz* (1879). *Werke und Briefe*, vol. 1. Edited by Fritz Bergemann. Frankfurt am Main: Insel Verlag, 1982.

Burger, Heinz Otto. *Die Gedankenwelt der großen Schwaben*. Stuttgart: J.F. Steinkopf Verlag, 1978.

Cassirer, Ernst. *An Essay on Man: An Introduction to a Philosophy of Human Culture*. New Haven: Yale University Press, 1944.

———. *Philosophie der symbolischen Formen. Zweiter Teil: Das mythische Denken*. Berlin: Bruno Cassirer, 1925. *The Philosophy of Symbolic Forms. Volume Two: Mythical Thought*. Translated by Ralph Manheim. New Haven: Yale University Press, 1955.

Cervantes Saavedra, Miguel de. *Don Quixote* (1605–1615). Translated by Edith Grossman. New York: Ecco, 2003.

Challiol-Gillet, Marie-Christine. *Schelling: une philosophie de l'extase*. Paris: Presses universitaires de France, 1998.

Coleridge, Samuel Taylor. *Biographia Literaria* (1817). Edited by James Engell and W. Jackson Bate. *The Collected Works of Samuel Taylor Coleridge*. Bollingen Series LXXV. Princeton: Princeton University Press, 1984.

Confucius. *The Analects of Confucius: A Philosophical Translation*. Translated by Roger T. Ames and Henry Rosemont Jr. New York: Random House, 2010.

Courtine, Jean-François, *Extase de la raison: Essais sur Schelling*. Paris: Galilée, 1990.
———. "Tragedy and Sublimity: The Speculative Interpretation of *Oedipus Rex* on the Threshold of German Idealism." *Of the Sublime: Presence in Question*. Translated by Jeffrey S. Librett. Albany: State University of New York Press, 1993, 157–174.
———, and Jean-François Marquet. *Le dernier Schelling: raison et positivité*. Paris: Librairie philosophique J. Vrin, 1994.
Cousin, Victor. *Über französische und deutsche Philosophie*. Stuttgart and Tübingen: J.G. Cotta, 1834.
Cuvier, Georges. *Georges Cuvier, Fossil Bones, and Geological Catastrophes*. Edited and translated by Martin J.S. Rudwick. Chicago and London: University of Chicago Press, 1997.
Danz, Christian. *Die philosophische Christologie F.W.J. Schellings*. Stuttgart (Bad Cannstatt): Frommann-Holzboog, 1996.
Das, Saitya Brata. *The Promise of Time: Towards a Phenomenology of Promise*. Shimla (Himachal Pradesh), India: Indian Institute of Advanced Study, 2011.
Dekker, Gerbrand Jan. *Die Rückwendung zum Mythos: Schellings letzte Wandlung*. Munich: Oldenbourg, 1930.
Deleuze, Gilles. *Cinema 2: The Time Image*. Translated by Hugh Tomlinson and Robert Galeta. Minneapolis: University of Minnesota Press, 1989.
———. *Différence et repetition*. Paris: Presses universitaires de France, 1968. *Difference and Repetition*. Translated by Paul Patton. New York: Columbia University Press, 1994.
———. *Expressionism in Philosophy: Spinoza* (1968). Translated by Martin Joughin. New York: Zone Books, 1992.
———. *The Fold: Leibniz and the Baroque* (1988). Translated by Tom Conley. Minneapolis: University of Minnesota Press, 1993.
———. *Francis Bacon: Logique de la sensation*. Paris: Éditions de la Différance, 1981. *Francis Bacon: The Logic of Sensation*. Translated by Daniel W. Smith. Minneapolis: University of Minnesota Press, 2003.
———. *Kant's Critical Philosophy* (1963). Translated by Hugh Tomlinson and Barbara Habberjam. Minneapolis: University of Minnesota Press, 1984.
———. *Nietzsche et la philosophie*. Paris: Presses universitaires de

France, 1962. *Nietzsche and Philosophy*. Translated by Hugh Tomlinson. New York: Columbia University Press, 1983.

———. *Spinoza: Practical Philosophy* (1970, revised 1981). Translated by Robert Hurley. San Francisco: City Lights Books, 1988.

———, and Félix Guattari. *Anti-Oedipus: Capitalism and Schizophrenia* (1972). Translated by Robert Hurley, Mark Seem, and Helen R. Lane. Minneapolis: University of Minnesota Press, 1983.

———, and Félix Guattari. *Qu'est-ce que la philosophie?* Paris: Les Éditions de Minuit, 1991. *What is Philosophy?* Translated by Hugh Tomlinson and Graham Burchell. New York: Columbia University Press, 1994.

———, and Félix Guattari. *A Thousand Plateaus*. Translated by Brian Massumi. Minneapolis: University of Minnesota Press, 1987.

Derrida, Jacques, *The Beast & the Sovereign*, vol. 1. Translated by Geoffrey Bennington. Chicago and London: University of Chicago Press, 2009.

Dōgen Eihei. *Shōbōgenzō*, 2 volumes. Edited Kazuaki Tanahashi. Boston and London: Shambhala, 2010.

Dostoevsky, Fyodor. *The Brothers Karamazov*. Translated by Richard Pevear and Larissa Volokhonsky. New York: FSG, 2002.

———. *Notes from Underground*. Translated by Richard Pevear and Larissa Volokhonsky. New York: Knopf, 1993.

Esposito, Joseph L. *Schelling's Idealism and Philosophy of Nature*. Lewisburg: Bucknell University Press, 1977.

Fackenheim, Emile L. *The God Within: Kant, Schelling, and Historicity*. Toronto: University of Toronto Press, 1996.

Fénelon, François, and Jeanne Guyon. *Spiritual Progress: Instructions in the Divine Life of the Soul*. Edited and translated by James W. Metcalf. New York: Dodd, 1853.

Ferreyra, Julián. "Flaubert's *Bêtise* as a Way into Deleuzian Ideas." Unpublished manuscript presented at Harvard University, February 16, 2012.

Fichte, Johann Gottlieb. *Fichtes Werke*, 11 volumes. Edited by Immanuel Hermann Fichte. Berlin: Walter de Gruyter, 1971.

Flaubert, Gustave. *Bouvard and Pécuchet*. Translated by T.W. Earp and G.W. Sonier. New York: New Directions, 1954.

———. *The Dictionary of Accepted Ideas*. Translated and edited by Jacques Barzun. Included as a supplement to *Bouvard and Pécuchet*.

Foucault, Michel. "Nietzsche, Genealogy, History." *The Foucault Reader*. Edited by Paul Rabinow. New York: Pantheon, 1984.

———. "Theatrum Philosophicum." Translated by Donald F. Brouchard and Sherry Simon. *Aesthetics, Method, and Epistemology: Essential Works of Foucault*, vol. 2. Edited by James D. Faubion. New York: New Press, 1998.

Frank, Manfred. *Eine Einführung in die frühromantische Ästhetik*. Frankfurt am Main: Suhrkamp, 1989.

———. *Der kommende Gott: Vorlesungen über die neue Mythologie*. Frankfurt am Main, 1982.

———. *Der unendliche Mangel an Sein: Schellings Hegel Kritik und die Anfänge der marxischen Dialektik*, 2nd, expanded edition. Munich: Wilhelm Fink, 1992.

———. *Eine Einführung in Schellings Philosophie*. Frankfurt am Main: Suhrkamp, 1985.

———. *The Philosophical Foundations of Early German Romanticism*. Translated by Elizabeth Millán-Zaibert. Albany: State University of New York Press, 2004.

———, and Kurz, Gerhard, eds. *Materialien zu Schellings philosophischen Anfängen*. Frankfurt am Main: Suhrkamp, 1975.

Franz, Albert. *Philosophische Religion: Eine Auseinandersetzung mit den Grundlegungsproblemen der Spätphilosophie Schellings*. Amsterdam: Rodopi, 1992.

Franz, Michael. *Schellings Tübinger Platon Studien*. Göttingen: Vandenhoeck and Ruprecht, 1996.

Freydberg, Bernard. *Schelling's Dialogical* Freedom Essay: *Provocative Philosophy Then and Now*. Albany: State University of New York Press, 2008.

Friedrich, Hans-Joachim. *Der Ungrund der Freiheit im Denken von Böhme, Schelling und Heidegger* (Schellingiana 24). Stuttgart (Bad Cannstatt): Frommann-Holzboog, 2009.

Fuhrmans, Horst. *Schellings letzte Philosophie: Die negative und positive Philosophie im Einsatz des Spätidealismus*. Berlin: Junker und Dünnhaupt Verlag, 1940.

———. *Schellings Philosophie der Weltalter: Schellings Philosophie in den Jahren 1806–1821*. Düsseldorf: L. Schwann, 1954.

Gabriel, Markus. *Das Absolute und die Welt in Schellings* Freiheitsschrift. Bonn: University Press, 2006.

———. *Der Mensch im Mythos: Untersuchungen über Ontotheologie,*

Anthropologie und Sebstbewußtseinsgeschichte in Schellings Philosophie der Mythologie. Berlin and New York: Walter de Gruyter, 2006.

———. *Transcendental Ontology: Essays in German Idealism*. London and New York: Continuum, 2011.

———, and Slavoj Žižek. *Mythology, Madness and Laughter: Subjectivity in German Idealism*. London and New York: Continuum, 2009.

Garcia, Marcela. "Schelling's Late Negative Philosophy: Crisis and Critique of Pure Reason." *Comparative and Continental Philosophy* 3.2 (Fall 2011), 141–163.

Gasché, Rodolphe. *Geophilosophy: On Gilles Deleuze and Félix Guattari's What is Philosophy?* Evanston: Northwestern University Press, 2014.

Gilman, Sander L. ed. *Conversations with Nietzsche: A Life in the Words of His Contemporaries*. Translated by David J. Parent. New York and Oxford: Oxford University Press, 1987.

Goudeli. Kyriaki. *Challenges to German Idealism: Schelling, Fichte and Kant*. Hampshire: Palgrave Macmillan, 2002.

Grant, Iain Hamilton. *Philosophies of Nature After Schelling*. London and New York: Continuum, 2006.

Groves, Christopher. "Ecstasy of Reason, Crisis of Reason: Schelling and Absolute Difference." *Pli* 8 (1999), 25–45.

Habermas, Jürgen. *Das Absolute und die Geschichte: Von der Zwiespältigkeit in Schellings Denken*. (Dissertation) Bonn: H. Bouvier, 1954.

Hablützel, Rudolf. *Dialektik und Einbildungskraft: F. W. J. Schellings Lehre von der menschlichen Erkenntnis*. Basel: Verlag für Recht und Gesellschaft, 1954.

Hamann, Johann Georg. *Aesthetica in Nuce: A Rhapsody in Cabbalistic Prose*. Translated by Joyce P. Crick, with modifications by H.B. Nisbet. *German Aesthetic and Literary Criticism: Winckelmann, Lessing, Hamann, Herder, Schiller, Goethe*. Edited by H.B. Nisbet. Cambridge: Cambridge University Press, 1985.

———. *Sämtliche Werke, volume III, Schriften über Sprache/Mysterien/Vernunft* (1772–1788). Edited by Josef Nadler. Vienna: Herder, 1951.

———. *Socratic Memorabilia* (1759), a bilingual edition. Translated by James C. O'Flaherty. Baltimore: Johns Hopkins Press, 1967.

Hegel, Georg Wilhelm Friedrich. *The Difference Between Fichte's and Schelling's System of Philosophy* (1801). Translated by H.S. Harris and Walter Cerf. Albany: State University of New York Press, 1977.

———. *Gesammelte Werke*. Edited by Hartmut Buchner and Otto Pöggeler. Hamburg: Felix Meiner, 1968.

———. *Hegel: The Letters*. Translated by Clark Butler and Christiane Seiler. Bloomington: Indiana University Press, 1984.

———. *Hegel's Aesthetics: Lectures on Fine Arts*, vol. 1. Translated by T.M. Knox. Oxford: Oxford University Press, 1975.

———. *Lectures on the History of Philosophy: The Lectures of 1825–1826: Volume Three: Medieval and Modern Philosophy*. Edited by Robert Brown. Translated by Robert Brown and J.M. Stewart with the assistance of H.S. Harris. Berkeley: University of California Press, 1990.

———. *Phenomenology of Spirit* (1807). Translated by A.V. Miller. Oxford: Oxford University Press, 1977.

———. *Werke in zwanzig Bänden: Theorie Werkausgabe*. Frankfurt am Main: Suhrkamp, 1971.

Heidegger, Martin. *Beiträge zur Philosophie (Vom Ereignis)*. Edited by Friedrich-Wilhelm von Herrmann. Frankfurt am Main: Vittorio Klostermann, 1989.

———. *Die Metaphysik des deutschen Idealismus (Schelling)*. Frankfurt: Vittorio Klostermann, 1991.

———. *Schellings Abhandlung über das Wesen der menschlichen Freiheit*. Edited by H. Feick. Tübingen: Max Niemeyer, 1971.

———. *Schelling's Treatise On the Essence of Human Freedom*. Translated by Joan Stambaugh. Athens: Ohio University Press, 1985.

Henningfeld, Jochem. *Mythos und Poesie: Interpretationen zu Schellings* Philosophie der Kunst *und* Philosophie der Mythologie. Meisenheim am Glan: Anton Hain, 1973.

Heuser-Kessler, Marie-Luise. *'Die Produktivität der Natur': Schellings Naturphilosophie und das neue Paradigma der Selbstorganisation in den Naturwissenschaften*. Berlin: Duncker & Humblot, 1986.

Höffe, Otfried, and Pieper, Annemarie, eds. *Über das Wesen der menschlichen Freiheit*. Vol. 3 of the *Klassiker Auslegen*. Berlin: Akademie Verlag, 1995.

Hogrebe, Wolfram. *Echo des Nichtwissens*. Berlin: Akademie Verlag, 2006.

———. *Prädikation und Genesis: Metaphysik als Fundamentalheuristik im Ausgang von Schellings* Die Weltalter. Frankfurt: Suhrkamp, 1989.

Hölderlin, Friedrich. *Essays and Letters on Theory*. Edited and translated by Thomas Pfau. Albany: State University of New York Press, 1988.

———. *Sämtliche Werke* (the so-called *Große Stuttgarter Ausgabe*). Stuttgart: W. Kohlhammer Verlag and J.G. Cottasche Buchhandlung Nachfolger, 1944.

Hühn, Lore. "Die anamnetische Historie des Anfangs: Ein Versuch zu Schelling und Kierkegaard." *Anfang und Ursprung: Die Frage nach dem Ersten in Philosophie und Kulturwissenschaft*. Edited by Emil Angehrn. Berlin and New York: de Gruyter, 2007, 203–213.

———. *Fichte und Schelling oder Über die Grenze menschlichen Wissens*. Stuttgart and Weimar: Verlag J. B. Metzler, 1994.

———. "Die Idee der neuen Mythologie: Schellings Weg einer naturphilosophischen Fundierung." *Evolution des Geistes—Jena um 1800: Natur und Kunst, Philosophie und Wissenschaft im Spannungsfeld der Geschichte*. Edited by Friedrich Strack. Stuttgart: Klett-Cotta, 1994, 393–411.

———. "Die intelligible Tat: Zu einer Gemeinsamkeit Schellings und Schopenhauers." *Selbstbesinnung der philosophischen Moderne: Beiträge zur kritischen Hermeneutik ihrer Grundbegriffe*. Edited by Christian Iber and Romano Pocai. Cuxhaven and Dartford: Junghans, 1998, 55–94.

———. *Kierkegaard und der deutsche Idealismus: Konstellationen des Übergangs*. Tübingen: Mohr Siebeck, 2009.

———. "A Philosophical Dialogue between Heidegger and Schelling." Translated by David Carus. *Comparative and Continental Philosophy* 6.1 (Spring 2014), 16–34.

———. "Der Wille, der Nichts will: Zum Paradox negativer Freiheit bei Schelling und Schopenhauer." In: *Die Ethik Arthur Schopenhauers im Kontext des Deutschen Idealismus (Fichte/Schelling)*. Edited by Lore Hühn. Würzburg: Ergon, 2006, 149–160.

———, editor. *Schopenhauer liest Schelling*. Stuttgart (Bad Cannstatt): Frommann-Holzboog, 2012.

———, and Jörg Jantzen (with the cooperation of Philipp Schwab and Sebastian Schwenzfeuer). *Heideggers Schelling-Seminar (1927/28)* (Schellingiana 22). Stuttgart (Bad Cannstatt): Frommann-Holzboog, 2010.

Hutter, Axel. *Geschichtliche Vernunft: Die Weiterführung der Kantischen Vernunftkritik in der Spätphilosophie Schellings*. Frankfurt am Main: Suhrkamp, 1996.

Jacobi, Friedrich Heinrich. *Werke. Gesamtausgabe volume three*: *Schriften zum Streit um die göttlichen Dinge und ihre Offenbarung*. Edited by

Klaus Hammacher und Walter Jaeschke. Hamburg: Felix Meiner Verlag, 2000.
Jacobs, W.G. *Gottesbegriff und Geschichtsphilosophie in der Sicht Schellings*. Stuttgart (Bad Cannstatt): Frommann-Holzboog, 1993.
Jähnig, Dieter. *Schelling: Die Kunst in der Philosophie*, 2 volumes. Pfullingen: Neske, 1966 and 1969.
———. *Der Weltbezug der Künste: Schelling, Nietzsche, Kant*. Freiburg: Verlag Karl Alber, 2011.
———. *Welt-Geschichte: Kunst-Geschichte: Zum Verhältnis von Vergangenheitserkenntnis und Veränderung*. Köln (Cologne): Dumont Schauberg, 1975.
Jaspers, Karl. *Schelling: Größe und Verhängnis* (1955). Munich: Piper, 1986.
———. "Schellings Größe und sein Verhängnis," *Studia Philosophica: Jahrbuch der Schweizerischen Philosophischen Gesellschaft*, vol. XIV. Basel: Verlag für Recht und Gesellschaft, 1954, 12–50.
Kahn-Wallerstein, Carmen. *Schellings Frauen: Caroline und Pauline*. Frankfurt am Main: Insel Verlag, 1979.
Kandinsky, Wassily. *Über das Geistige in der Kunst* (1911). Bern: Benteli Verlag, 1952.
Kant, Immanuel. *Anthropology, History, and Education*. Edited and translated by Robert B. Louden and Günter Zöller. Cambridge: Cambridge University Press, 2008.
———. *The Conflict of the Faculties* [*Der Streit der Fakultäten*] (1798). Translated by Mary J. Gregor. Lincoln: University of Nebraska Press, 1992.
———. *Kant's Gesammelte Schriften*. 29 volumes. Berlin: Berlin Akademie, 1900– [the so-called *Akademie Ausgabe*].
———. *Kritik der reinen Vernunft* (1781). Edited by Raymund Schmidt. Hamburg: Felix Meiner Verlag, 1956. *Critique of Pure Reason*. Translated by Norman Kemp Smith. New York: St. Martin's Press, 1965.
———. *Kritik der Urteilskraft* (1790). Edited by Karl Vorländer. Hamburg: Felix Meiner Verlag, 1990. *Critique of Judgment*. Translated by Werner S. Pluhar. Indianapolis: Hackett, 1987.
———. *Opus Postumum*. Edited by Eckart Förster. Translated by Eckart Förster and Michael Rosen. Cambridge: Cambridge University Press, 1993.
———. *Religion Within the Limits of Reason Alone* (1793). Translated by Theodore M. Greene and Hoyt H. Hudson. New York: Harper Torchbooks, 1960.

———. *Werke*. Edited by Ernst Cassirer. Berlin: Bruno Cassirer Verlag, 1923.

Kasper, Walter. *Das Absolute in der Geschichte: Philosophie und Theologie der Geschichte in der Spätphilosophie Schellings*. Mainz: Matthias-Grünewald, 1965.

Kerslake, Christian. *Immanence and the Vertigo of Philosophy: From Kant to Deleuze*. Edinburgh: Edinburgh University Press, 2009.

Kolbert, Elizabeth. "The Sixth Extinction?" *New Yorker*, May 25, 2009.

———. *The Sixth Extinction: An Unnatural History*. New York: Holt, 2014.

Krell, David Farrell. "The Crisis of Reason in the Nineteenth Century: Schelling's Treatise 'On Human Freedom' (1809)." *The Collegium Phaenomenologicum: The First Ten Years*. Edited by John Sallis, Giuseppina Moneta, and Jacques Taminiaux. Dordrecht: Kluwer, 13–32.

———. *The Tragic Absolute: German Idealism and the Languishing of God*. Bloomington and Indianapolis: Indiana University Press, 2005.

Krings, Hermann. "Das Prinzip Existenz in Schellings Weltaltern." *Symposium* 4 (1955), 335–347.

Kundera, Milan. *Farewell Waltz*. Translated by Aaron Asher. New York: HarperCollins, 1998.

———. *The Unbearable Lightness of Being*. Translated by Michael Heim. New York: Harper & Row, 1984.

Küppers, Bernd-Olaf. *Natur als Organismus: Schellings frühe Naturphilosophie und ihre Bedeutung für die moderne Biologie*. Frankfurt am Main: Vittorio Klostermann, 1992.

Lacan, Jacques. "The Mirror Stage as Formative of the Function of the I as Revealed in Psychoanalytic Experience" (1949). *Écrits: A Selection*. Translated by Alan Sheridan. New York and London: Norton, 1977.

Lauer, Christopher. *The Suspension of Reason in Hegel and Schelling*. London and New York: Continuum, 2010.

Laughland, John. *Schelling versus Hegel: From German Idealism to Christian Metaphysics*. Aldershot, Hampshire and Burlington, VT: Ashgate, 2007.

Lawrence, Joseph P. *Schellings Philosophie des ewigen Anfangs: Die Natur als Quelle der Geschichte*. Würzburg: Königshausen und Neumann, 1989.

Leakey, Richard, and Roger Lewin. *The Sixth Extinction: Patterns of Life and the Future of Humankind*. New York: Doubleday, 1995.

Leopold, Aldo. *A Sand County Almanac and Sketches Here and There* (1949). Oxford: Oxford University Press, 1968.

Loer, Barbara. *Das Absolute und die Wirklichkeit in Schellings Philosophie: Mit der Erstedition einer Handschrift aus dem Berliner Schelling-Nachlass*. Berlin: De Gruyter, 1974.

Löwith, Karl. *Nietzsches Philosophie der ewigen Wiederkunft des Gleichen*. Berlin: Die Runde, 1955.

Lukács, Georg. *Die Zerstörung der Vernunft: Der Weg des Irrationalismus von Schelling zu Hitler*. Berlin: Aufbau Verlag, 1955.

Lyotard, Jean-François. *The Differend: Phrases in Dispute*. Translated by Georges van den Abbeele. Minneapolis: University of Minnesota Press, 1989.

Malabou, Catherine. *Plasticity at the Dusk of Writing: Dialectic, Destruction, Deconstruction*. Translated by Carolyn Shread. New York: Columbia University Press, 2010.

Marcel, Gabriel. *Coleridge et Schelling*. Paris: Aubier-Montaigne, 1971.

Marks, Ralph. *Konzeption einer dynamischen Naturphilosophie bei Schelling und Eschenmayer*. Munich: Holler, 1982.

Marovitz, Sanford. "Melville's Problematic 'Being.'" *Bloom's Critical Modern Views: Melville*, new edition. Edited by Harold Bloom. New York: Infobase, 2008.

Marquet, Jean-François. *Liberté et existence: Étude sur la formation de la philosophie de Schelling*. Paris: Gallimard, 1973.

Marquard, Odo. *Farewell to Matters of Principle: Philosophical Studies*. Translated by Robert M. Wallace with Susan Bernstein and James I. Porter. New and Oxford: Oxford University Press, 1989.

———. *Transzendentaler Idealismus, Romantische Naturphilosophie, Psychoanalyse*. Cologne: Verlag für Philosophie Jürgen Dinter, 1987. (This was his 1962 *Habilitationsschrift*.)

Marx, Werner. *Schelling: Geschichte, System, Freiheit*. Freiburg and Munich: Karl Alber, 1977. *The Philosophy of F. W. J. Schelling: History, System, and Freedom*. Translated by Thomas Nenon. Bloomington: Indiana University Press, 1984.

Matthews, Bruce. "Rationality's Demand of its Other: A Comparative Analysis of F.W.J. Schelling's *Unvordenkliche* and Huineng's *Wu-Nien*." *Comparative and Continental Philosophy* 4.1 (Spring 2012), 75–92.

———. *Schelling's Organic Form of Philosophy: Life as the Schema of Freedom*. Albany: State University of New York Press, 2011.

Maturana, Humberto, and Francisco Varela. *Autopoiesis and Cognition: The Realization of the Living*. Dordecht: Reidel, 1980.

May, Todd. *Gilles Deleuze: An Introduction*. Cambridge: Cambridge University Press, 2005.

McCumber, John. *On Philosophy: Notes from a Crisis*. Stanford: Stanford University Press, 2013.

McFarland, Thomas. *Coleridge and the Pantheist Tradition*. Oxford: Clarendon, 1969.

McGrath, Sean. *The Dark-Ground of Spirit: Schelling and the Unconscious*. London and New York: Routledge, 2011.

———. "Schelling on the Unconscious." *Research in Phenomenology* 40.1, 72–91.

Melville, Herman. *The Letters of Herman Melville*. Edited by Merrell R. Davis and William H. Gilman. New Haven: Yale University Press, 1960.

———. *Moby-Dick: Or, the Whale* (1851). The Northwestern-Newberry Edition of the Writings of Herman Melville, vol. 6. Edited by Harrison Hayford, Hershel Parker, and G. Thomas Tanselle. Evanston and Chicago: Northwestern University Press and the Newberry Library, 1988.

Merleau-Ponty, Maurice. *Nature: Course Notes from the Collège de France*. Edited by Dominique Séglard. Translated by Robert Vallier. Evanston: Northwestern University Press, 2003.

Musil, Robert. *The Man Without Qualities*, vol. 1. Translated by Sophie Wilkins. New York: Vintage, 1996.

———. "On Stupidity" (1937). *Precision and Soul*. Translated by Burton Pike and David S. Luft. Chicago and London: University of Chicago Press, 1990.

Nancy, Jean-Luc. *Being Singular Plural*. Translated by Robert D. Richardson and Anne E. O'Byrne. Stanford: Stanford University Press, 2000.

———. *The Muses*. Translated by Peggy Kamuf. Stanford: Stanford University Press, 1996.

Newman, Barnett. "The New Sense of Fate." *Barnett Newman: Selected Writings and Interviews*. Edited by John P. O'Neill. Berkeley: University of California Press, 1990.

Nietzsche, Friedrich, *Kritische Gesamtausgabe*. Edited by Giorgio Colli

and Mazzino Montinari. Berlin and New York: Walter de Gruyter, 1967ff.
———. *Sämtliche Werke: Kritische Studienausgabe in 15 Einzelbänden*. Edited by Giorgio Colli and Mazzino Montinari. Munich: Deutscher Taschenbuch Verlag and Berlin: Walter de Gruyter, 1988.
———. *Werke in drei Bänden* (the so-called *Schlechta-Ausgabe*). Edited by Karl Schlechta. Munich: Hanser, 1959.
Nishida, Kitarō. *Art and Morality* (1923). Translated by David Dilworth and Valdo Viglielmo. Honolulu: University Press of Hawaii, 1973.
———. *An Inquiry into the Good* (1911). Translated by Masao Abe and Christopher Ives. New Haven and London: Yale University Press, 1990.
———. *Intelligibility and the Philosophy of Nothingness*. Translated by Robert Schinzinger. Honolulu: East-West Center Press, 1966.
———. *Last Writings: Nothingness and the Religious Worldview* (1945). Translated by David Dilworth. Honolulu: University Press of Hawaii, 1987.
Nishitani, Keiji. *Religion and Nothingness*. Translated by Jan van Bragt. Berkeley: University of California Press, 1982.
Norman, Judith and Alistair Welchman, eds. *The New Schelling*. London and New York: Continuum, 2004.
Ōhashi Ryōsuke. *Ekstase und Gelassenheit: Zu Schelling und Heidegger*. Munich: Wilhelm Fink, 1975.
Olson, Charles. *Call Me Ishmael* (1947). Baltimore and London: Johns Hopkins University Press, 1997.
O'Meara, Thomas F. *Romantic Idealism and Roman Catholicism: Schelling and the Theologians*. South Bend, Indiana, and London: University of Notre Dame Press, 1982.
Orsini, Gian N.G. *Coleridge and German Idealism: A Study in the History of Philosophy*. Carbondale and Edwardsville: Southern Illinois University Press, 1969.
Outram, Dorinda. *Georges Cuvier: Vocation, Science, and Authority in Post-Revolutionary France*. Manchester: Manchester University Press, 1984.
Pinkard, Terry. *German Philosophy 1760–1860: The Legacy of Idealism*. Cambridge: Cambridge University Press, 2002.
———. *Hegel: A Biography*. Cambridge: Cambridge University Press, 2000.

Redding, Paul. *Continental Idealism: Leibniz to Nietzsche*. London and New York: Routledge, 2009.

Richards, Robert J. *The Romantic Conception of Life: Science and Philosophy in the Age of Goethe*. Chicago and London: University of Chicago Press, 2002.

Rolston III, Holmes. "Challenges in Environmental Ethics." *Environmental Philosophy: From Animal Rights to Radical Ecology*, 4th edition. Edited by Michael E. Zimmerman, J. Baird Callicott, Karen J. Warren, Irene J. Klaver, and John Clark. Upper Saddle River, NJ: Pearson, 2005.

Ronell, Avital. *Stupidity*. Urbana and Chicago: University of Illinois Press, 2002.

Ruddiman, William F. *Plows, Plagues, and Petroleum: How Humans Took Control of Climate*. Princeton: Princeton University Press, 2007.

Sallis, John. *Chorology: On Beginning in Plato's* Timaeus. Bloomington: Indiana University Press, 1999.

———. *Force of Imagination: The Sense of the Elemental*. Bloomington: Indiana University Press, 2000.

———. *Logic of Imagination: The Expanse of the Elemental*. Bloomington and Indianapolis: Indiana University Press, 2012.

———. *Transfigurements: On the True Sense of Art*. Chicago and London: University of Chicago Press, 2008.

Sandkühler, Hans-Jörg. *Freiheit und Wirklichkeit: Zur Dialektik von Politik und Philosophie bei Schelling*. Frankfurt am Main: Suhrkamp, 1968.

———. *Friedrich Wilhelm Joseph Schelling*. Stuttgart: Metzler, 1970.

Schraven, Martin. *Philosophie und Revolution: Schellings Verhältnis zum Politischen im Revolutionsjahr 1848*. Stuttgart (Bad Cannstatt): Frommann-Holzboog, 1989.

Schröter, Manfred. *Über Schelling und zur Kulturphilosophie*. Munich: Oldenbourg, 1971.

Schulte, Christoph. "Zimzum bei Schelling." *Kabbala und Romantik: Die jüdische Mystik in der romantischen Geistesgeschichte*. (*Conditio Jadaica* 7.) Edited by Eveline Goodman-Thau, Gert Mattenklott, and Christoph Schulte. Tübingen: Max Niemeyer, 1994, 97–118.

Schulz, Walter. *Die Vollendung des deutschen Idealismus in der Spätphilosophie Schellings*. 2nd edition. Pfullingen: Neske, 1975.

Shaw, Devin Zane. *Freedom and Nature in Schelling's Philosophy of Art*. London and New York: Continuum, 2010.

Singh, Rustam. "*Weeping*" *and Other Essays on Being and Writing*. Jaipur, India: Pratilipi Books, 2011.

Snow, Dale E. *Schelling and the End of Idealism*. Albany: State University of New York Press, 1996.

Snyder, Gary. *A Place in Space: Ethics, Aesthetics, and Watersheds*. Washington, DC: Counterpoint, 1995.

———. *The Practice of the Wild*. San Francisco: North Point Press, 1990.

Stevens, Wallace. *The Necessary Angel: Essays on Reality and the Imagination* (1951). *Collected Poetry and Prose*. Edited by Frank Kermode and Joan Richardson. New York: Library of America, 1997.

Summerell, Orrin F. "The Theory of the Imagination in Schelling's Philosophy of Identity." *Idealistic Studies* 34.1 (2004), 85–98.

Sun Qianli. "Treatise on Calligraphy" (687). *Two Chinese Treatises on Calligraphy*. Translated by Chang Ch'ung and Hans H. Frankel. New Haven and London: Yale University Press, 1995.

Thomas. William. *The Finitudes of God: Notes on Schelling's Handwritten Remains*. San Jose, New York, Lincoln, Shanghai: Writers Club Press, 2002.

Tillich, Paul. *Mystik und Schuldbewußtsein in Schellings philosophischer Entwicklung*. In *Gesammelte Werke 1*. Stuttgart: Evangelisches Verlagswerk, 1959, 13–108. *Mysticism and Guilt-Consciousness in Schelling's Philosophical Development*. Translated by Victor Nuovo. Lewisburg: Bucknell University Press, 1974.

Tilliette, Xavier. *L'absolu et la philosophie: Essais sur Schelling*. Paris: Presses universitaires de France, 1987.

———. *La mythologie comprise: L'interpretation Schellingienne du paganisme*. Naples: Bibliopolis, 1984.

———. *Schelling: Biographie*. Paris: Calmann-Lévy, 1999.

———. *Schelling: une philosophie en devenir, two volumes*. Paris: J. Vrin, 1970.

Toscano Alberto. "Fanaticism and Production: On Schelling's Philosophy of Indifference." *Pli* 8 (1999): 46–70.

———. "Philosophy and the Experience of Construction." *The New Schelling*. Edited by Judith Norman and Alistair Welchman. London and New York: Contiuum, 2004, 106–127.

Tritten, Tyler. *Beyond Presence: The Late F. W. J. Schelling's Criticism of Metaphysics*. Boston and Berlin: De Gruyter, 2012.

Vattimo, Gianni. *After Christianity*. Translated by Luca D'Isanto. New York: Columbia University Press, 2002.

The Vimalakīrti Sutra. Translated by Burton Watson. New York: Columbia University Press, 1997.

Walsh, David. *The Modern Philosophical Revolution: The Luminosity of*

Existence. Cambridge and New York: Cambridge University Press, 2008.

Whistler, Daniel. "Language after Philosophy of Nature: Schelling's Geology of Divine Names." *After the Postsecular and the Postmodern: New Essays in Continental Philosophy of Religion*. Edited by Anthony Paul Smith and Daniel Whistler. Newcastle: Cambridge Scholars' Press, 2010, 335–359.

———. *Schelling's Theory of Symbolic Language: Forming the System of Identity*. Oxford: Oxford University Press, 2013.

Wieland, Wolfgang. *Schellings Lehre von der Zeit: Grundlagen und Voraussetzungen der Weltalterphilosophie*. Heidelberg: Carl Winter Universitätsverlag, 1956.

Wild, Christoph. *Reflexion und Erfahrung: Eine Interpretation der Früh- und Spätphilosophie Schellings*. Freiburg and Munich: Karl Alber, 1968.

Wilson, John Elbert. *Schelling und Nietzsche: Zur Auslegung der frühen Werke Friedrich Nietzsches*. Berlin and New York: Walter de Gruyter, 1996.

———. *Schellings Mythologie: Zur Auslegung der Philosophie der Mythologie und der Offenbarung*. Stuttgart (Bad Cannstatt): Frommann-Holzboog, 1993.

Wirth, Jason. "The Abject Root: Kant and the Problem of Representing Evil," *Kant: Making Reason Intuitive*. Edited by Patellis, Goudeli, and Kontos. Hampshire: Palgrave Macmillan, 2007, 146–163.

———. *Commiserating with Devastated Things: Milan Kundera and the Entitlements of Thought*. New York: Fordham University Press, 2015.

———. *The Conspiracy of Life: Meditations on Schelling and His Time*. Albany: State University of New York Press, 2003.

———. "Dōgen and the Unknown Knowns: The Practice of the Wild After the End of Nature." *Environmental Philosophy* 10:1 (Spring 2013), 39–62.

———. "The Language of Natural Silence: Schelling and the Poetic Word After Kant." *The Linguistic Dimension of Kant's Thought: Historical and Critical Essays*. Edited by Frank Schalow and Richard Velkley. Evanston: Northwestern University Press, 2014, 237–262.

———. "*Mitwissenschaft*: Schelling and the Ethical." *Epochē: A Journal for the History of Philosophy* 8:2 (Spring 2004), 215–232.

———. "One Bright Pearl: On Japanese Aesthetic Expressivity." *The Movement of Nothingness: Trust in the Emptiness of Time*. Edited by Daniel Price and Ryan Johnson. Aurora, CO: Davies, 2013, 21–36.

———. "The Return of the Repressed: Schelling, Kierkegaard, and *Nachträglichkeit* in the Legacy of German Idealism." *Research in Phenomenology*, vol. 41 (2011).
———. "Schelling and the Force of Nature." *Interrogating the Tradition*. Edited by John Sallis and Charles Scott. Albany: State University of New York Press, 2000, 255–274.
———. "Schelling and the Future of God," *Analecta Hermeneutica*, vol. 5 (2014).
———. "Schelling's Contemporary Resurgence: The Dawn After the Night When All Cows Were Black." *Philosophy Compass* 6/9 (2011), 585–598.
———, editor (with Patrick Burke). *The Barbarian Principle: Merleau-Ponty, Schelling, and the Question of Nature*. Albany: State University of New York, 2013.
———, editor. *Schelling Now: Contemporary Readings*. Bloomington: Indiana University Press, 2004.
Wood, David. *The Step Back: Ethics and Politics after Deconstruction*. Albany: State University of New York Press, 2005.
Yates, Christopher. *The Poetic Imagination in Heidegger and Schelling*. London and New York: Bloomsbury Academic, 2013.
Žižek, Slavoj. *In Defense of Lost Causes*. London and New York: Verso, 2008.
———. *The Indivisible Remainder: An Essay on Schelling and Related Matters*. London and New York: Verso, 1996.

INDEX

GREEK TERMS